...dents and External...

Di...

CONTROL MECHANISMS IN DEVELOPMENT

**Activation, Differentiation, and
Modulation in Biological Systems**

ADVANCES IN EXPERIMENTAL MEDICINE AND BIOLOGY

Recent Volumes in this Series

CONTROL MECHANISMS IN DEVELOPMENT

Activation, Differentiation, and
Modulation in Biological Systems

Edited by

Russel H. Meints
and Eric Davies

Section of Cell Biology and Genetics
School of Life Sciences
University of Nebraska
Lincoln, Nebraska

PLENUM PRESS • NEW YORK AND LONDON

Library of Congress Cataloging in Publication Data
Main entry under title:

Control mechanisms in development.

   (Advances in experimental medicine and biology ; v. 62)
   Sponsored by the University of Nebraska.
   Includes bibliographies.
   1. Developmental biology—Congresses. 2. Biological control systems—
Congresses. I. Meints, Russel H. II. Davies, Eric. III. Nebraska. University.
School of Life Sciences. IV. Nebraska. University. V. Series. [DNLM: 1.
Cytogenetics—Congresses. AD 559 v.62 1974 / QH604 C764 1974]
QH491.C66   574.1'8   75-28152
ISBN 0-306-39062-0

Proceedings of a Symposium to inaugurate the School of Life Sciences
at the University of Nebraska-Lincoln, held October 14-16, 1974

© 1975 Plenum Press, New York
A Division of Plenum Publishing Corporation
227 West 17th Street, New York, N.Y. 10011

United Kingdom edition published by Plenum Press, London
A Division of Plenum Publishing Company, Ltd.
Davis House (4th Floor), 8 Scrubs Lane, Harlesden, London, NW10 6SE, England

Printed in the United States of America

To our wives and children with gratitude and affection

R.H.M.
E.D.

ACKNOWLEDGMENTS

The Symposium Committee would like to thank the following for their generous support:

College of Arts & Sciences, Mel George, Dean
School of Life Sciences, J.M. Daly, Interim Director
The University of Nebraska Foundation
Faculty Convocation Committee
Sigma Xi

R.H. Meints
Eric Davies
Jeffrey Hazel
James Van Etten
John Brumbaugh
J.M. Daly
John Janovy Jr.

PREFACE

This symposium was not only a happy event for the University of Nebraska, but it marked a milestone in the history of the biological sciences here. The symposium celebrated, in the most appropriate way possible, the creation of the new School of Life Sciences and ushered in what I believe will be a period of substantial development for biology on this campus. I am immensely proud of the faculty of this new School, and I have every confidence that the School's reputation and achievements will continue to grow.

As you all know, this university has had and still has distinguished scientists in the biological sciences and has offered fine programs at both the undergraduate and graduate level. But both the formation of the School of Life Sciences and the construction of the new Life Sciences Building promise a brighter future in this important area.

The School of Life Sciences was formed from the Departments of Botany, Microbiology, and Zoology, together with staff members in Biochemistry (from both the Department of Chemistry and from the former Department of Biochemistry and Nutrition in the College of Agriculture) as well as staff members in the College of Agriculture's Department of Plant Pathology. Our whole notion was to build a core unit in biology that would cross the lines between the College of Arts and Sciences and the College of Agriculture in order to combine strengths which exist in both areas. Despite the administrative difficulties that could have stood in the way of this development, it has proved to be a very workable concept, and we are delighted at the way things are going.

Why should the university be building additional strengths in biology at this time? First of all, we all recognize that agriculture is of vital importance not only to Nebraska but to the entire world. Advances in agriculture are clearly going to depend on a solid foundation of basic biological research. Secondly, biology is obviously deeply involved in questions of environment, of health care, of aging and of other societal problems. This is not in any way to suggest that an understanding of biology for its own sake is not important, but to point out some of the areas in which the university feels it has an obligation to contribute and in which strength in basic and applied biology is absolutely essential. With the combination of the strong College of Agriculture, a strong School of Life Sciences, and continued commitment on the part of the university and the state, we believe we have a magnificent opportunity to build a center of excellence in biology that will incorporate the excitement of this field into undergraduate education while training future scientists and pushing back the frontiers of knowledge.

As a personal statement, I'd like to express one hope—which applies not only to the future development of our School of Life Sciences, but to biology more generally—and that is that the ethical and value questions which are raised by advances in biological research become a concern of all biological scientists and a part of the curriculum of both undergraduate and graduate programs in biology. I feel strongly, for example, that a young person who wants training in biology to go into one of the health professions needs to be faced squarely with the issues involved in deciding when to turn off the machines and let a terminally ill patient die in dignity. Any student studying ecology and the environment needs to be faced with the economic and political issues involved in a decision as to whether or not the coal resources of Wyoming will be exploited.

We live in a world in which issues are becoming more intertwined and I believe that it is no longer possible nor proper for a scientist, perhaps particularly in biology, to be trained in isolation from the questions of value, of politics, and of economics that are becoming inevitably associated with advances in the field. I think it is not enough for scientists simply to say that such things are taken care of by requirements that force students to take some courses in the humanities and social sciences. Rather, I think the scientist must initiate some dialogue with his colleagues in those fields to try to insure that educational experiences for students are structured in such a way as to insure examination of such questions. It is my hope that the bright future which I see for the School of Life Sciences and for biology at the University of Nebraska will include not only excellence in research but also a commitment to teaching programs which will make such concerns a central part of the educational experience.

It's a great privilege for the University of Nebraska to have sponsored this inaugural symposium. We hope that it will contribute to the advancement of knowledge in biology even as it has stimulated us to continue our efforts in teaching and research in the life sciences.

Melvin D. George
Dean, College of Arts and Sciences
October 14, 1974

CONTENTS

ACTIVATION IN BIOLOGICAL SYSTEMS

---

*Denotes Symposium Participant.

DIFFERENTIATION IN BIOLOGICAL SYSTEMS

*J.E. Varner
Biology Department
Washington University
St. Louis, MO

*Robert T. Schimke, David J. Shapiro and G. Stanley McKnight
The Department of Biological Sciences
Stanford University
Stanford, CA

*Fotis C. Kafatos
The Biological Laboratories
Harvard University
Cambridge, MA

MODULATION IN BIOLOGICAL SYSTEMS

Yuzuru Akamatsu, Peter E. Dunn, Ferenc J. Kezdy,
Karl J. Kramer, *John H. Law, David Reibstein
and Larry L. Sanburg
Department of Biochemistry
University of Chicago
Chicago, IL

*David D. Sabatini, George Ojakian, Mauricio A. Lande,
John Lewis, Winnie Mok, Milton Adesnik, Gert Kreibich
Department of Cell Biology
New York University School of Medicine
New York, NY

CONTENTS

*S.J. Singer
Department of Biology
University of California at San Diego
La Jolla, CA

*Daniel E. Atkinson
Biochemistry Division
Department of Chemistry
University of California
Los Angeles, CA

PHILOSOPHICAL IMPLICATIONS

*Theodore T. Puck
Department of Biophysics and Genetics
University of Colorado Medical Center
Denver, CO

PREFORMED mRNA AND THE PROGRAMMING OF EARLY EMBRYO DEVELOPMENT

Abraham Marcus, Sara Spiegel and John D. Brooker

The Institute for Cancer Research, Fox Chase Center
for Cancer and Medical Sciences, Philadelphia, Pennsylvania

Introduction

Growth and development of an organism can be characterized as the implementation of a series of sequentially operative genetic programs. The process includes first a determination of the particular program, *i.e.* selectivity at the level of genomic transcription, and then either an immediate proceeding into its application, *i.e.* translation, or alternatively, holding the program in abeyance awaiting an activation reaction. In two systems, both of which involve early embryonic development, there appears to be substantial evidence that the latter of these two alternatives, namely the activation of a previously transcribed program, does indeed serve as a significant control. The systems are the fertilization of echinoderm and amphibian eggs, and the germination of seed embryos.

Protein Synthesis in Early Sea Urchin Development

Studies with sea urchin eggs have been primarily responsible for the concept of "maternal mRNA" as an important functioning entity in early embryonic development. Shortly after fertilization of the sea urchin egg, there occurs a striking increase in the rate of protein synthesis (Fig. 1) (Hultin, 1961; Nakana and Monroy, 1958; Epel, 1967). Initially it was considered that this increase might be due to an impairment in the translational machinery of the unfertilized egg, with fertilization setting into motion processes to correct the translational deficiency. Specifically, the suggestions included inadequate tRNAs (Ceccarina *et al.,* 1967), inactive ribosomes (Monroy *et al.,* 1965), and inhibitors

---

*Symposium Participant.

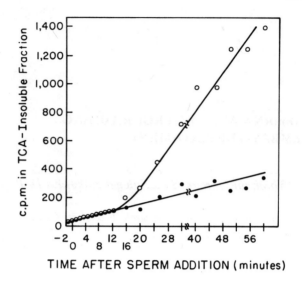

Figure 1. Cumulative Incorporation of ^{14}C-leucine by Unfertilized (dots) and Fertilized (circles) Eggs. From D. Epel (1967).

of protein synthesis (Metafora *et al.*, 1971; Gambino *et al.*, 1973). *In vitro* tests of these possibilities (Stavy and Gross, 1967), however, together with quantitative consideration of *in vivo* data (Humphreys, 1969), have led to the conclusion that the impairment mechanism could at most account for only a small part of the increase in protein synthesis.

An alternative possibility is suggested by the fact that the polyribosome content of sea urchin eggs increases markedly upon fertilization (Monroy and Tayler, 1963; Hultin, 1964; Stafford *et al.*, 1964; Rinaldi and Monroy, 1969). Such an observation suggests that the rate-limiting step in early protein synthesis is the mobilization of mRNA. This mobilization of mRNA, however, appears not to be a consequence of transcription of nuclear DNA. When actinomycin D is added during fertilization, at a level sufficient to inhibit greater than 90% of new RNA synthesis, there is little effect on early development (Gross *et al.*, 1964). Similarly, when enucleated eggs are activated parthenogenetically, early development occurs normally with the concomitant increased rate of protein synthesis (Denny and Tyler, 1964). Clearly, mRNAs that are transcribed during the early hours after fertilization are not obligatory to the initial developmental processes. Rather, it is the "maternal mRNAs" that have been transcribed during oogenesis that are mobilized into polyribosomes.

Two further lines of evidence support the "maternal RNA" concept. In DNA hybridization experiments (Whitely *et al.*, 1966; Davidson and Hough, 1971), oocyte RNA is found to hybridize to a fraction of cellular DNA considerably in excess of the content of ribosomal and transfer DNA. Similarly, in studies of amino acid incorporation, oocyte RNA is found to have considerable template activity (Slater and Spiegelman, 1966; Jenkins *et al.,* 1973).

Does the phenomenon of utilization of "preformed mRNA" extend to other biological systems? In the case of amphibian oogenesis, it has been reported that considerable mRNA [assayed either by template activity (Davidson *et al.,* 1966) or by poly U hybridization (Rosbash and Ford, 1974)] is made beginning at the lampbrush stage of the oocyte and that this mRNA is in large part retained in the mature egg (Davidson *et al.,* 1966; Rosbash and Ford, 1974). Furthermore, enucleated frog eggs have the same increased rate of protein synthesis after activation as do normal nucleated eggs (Fig. 2) (Ecker *et al.,* 1968). The amphibian situation differs somewhat from that of echinoderm oogenesis in that the increased rate of protein synthesis in the amphibian is initiated as the last stage in oocyte maturation (Smith *et al.,* 1966). Nevertheless, with regard to the delayed utilization of preformed mRNA for protein synthesis, the two systems are analogous.

Somewhat similar results, though considerably more incomplete, have been reported for several other systems. In the clam, *Spisula,* fertilization increases the rate of protein synthesis 4-fold in a process unaffected by prevention of RNA synthesis (Bell and Reeder, 1967). In the teleost fish, *Fundulus,* information for early development comes from mRNA previously synthesized in the ovuum (Crawford *et al.,* 1973). In germinating

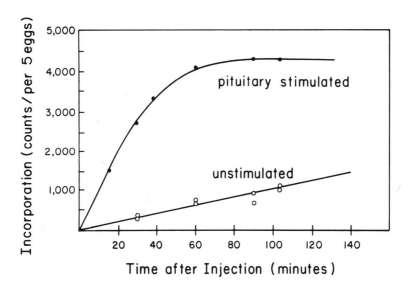

Figure 2. Kinetics of Incorporation of Tritated L-leucine into Stimulated and Unstimulated Rana pipiens Oocytes. From R.E. Ecker, L.D. Smith, and S. Subtelny (1968).

spores of the fungus *Botryodiplodia theobromae* there occurs increased protein synthesis supported by a preexistent mRNA (Brambl and Van Etten, 1970), while germination of lily pollen results in considerable augmenting of the polyribosome content accompanied by the initiation of protein synthesis (Mascarenhas and Bell, 1966; Linskins, 1969). In two additional systems, both of plant seed origin, the relationship between preformed mRNA and protein synthesis has been investigated in greater detail.

The Role of Preformed mRNAs in Seed Germination

Germination of seed embryos can be visualized by the change in fresh weight upon exposure to water (Marcus *et al.*, 1966; Pollock and Toole, 1966). Figure 3A presents a typical analysis for wheat embryos with an initially rapid increase to a value somewhat greater than twice the initial embryo weight, a quiescent period of about 5.5 hours during which time there is no change in fresh weight, and finally a slow sustained secondary increase in fresh weight. The initial increase due to imbibition occurs in nonviable as well as in viable embryos, while the secondary increase in fresh weight occurs only with viable embryos. One of the earliest changes occurring as a consequence of imbibition is a striking increase in the polyribosome content and the amino acid incorporating capacity of the embryos. The quantitative aspects of the change in amino acid incorporating

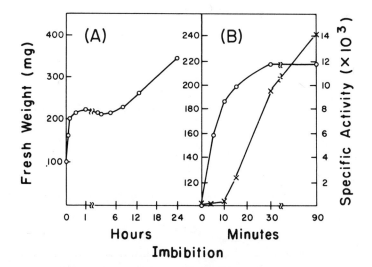

Figure 3. A) Time Course of Water Uptake by Wheat Embryos. One hundred mg samples of embryos were imbibed for the times shown, blotted thoroughly and dried. B) Comparison of water uptake and ribosomal activity in the first phase of imbibition. One hundred mg samples of embryos were analyzed for fresh weight as in A) and the ribosomes were isolated and assayed for activity in amino acid incorporation (see Marcus *et al.*, 1966; Marcus and Feeley, 1966).

capacity are best determined by isolating ribosomes from embryos at various times after exposure to water and assaying their activity in an *in vitro* system (Fig. 3B). Changes in polyribosome content can be determined by collecting the areas under the monoribosome and polyribosome peaks and determining the relative ribosome content (Fig. 4). Based on these types of experiments, we have concluded that the major increase in polyribosome formation occurs within the first 40 minutes after exposure to water. With some embryo preparations, there is little further increase in polyribosome content through the first 6 hours of germination, while with other preparations we have found further increases of up to 15%. Evidence that the increase in protein synthetic capacity is obligatory to germination is shown in Figure 5. Presence of 5 μg/ml cycloheximide in the imbibition solution inhibits the secondary increase in embryo fresh weight by greater than 90% at 16.5 hours and by about 85% at 22 hours. If the embryos originally imbibed in cycloheximide are washed at 5.5 hours, there is an almost complete recovery of the capacity for gain in fresh weight. The recovery includes, however, a 5.5-hour quiescent period similar to that observed with untreated embryos, thus suggesting that a period of protein synthesis is an absolute requirement prior to the resumption of embryo growth.

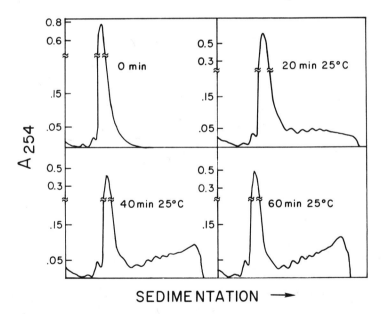

SEDIMENTATION ⟶

Figure 4. Time Course of Polyribosome Formation Upon Exposure to Water. Embryos were imbibed at 25° for the time periods shown, ribosomes isolated and fractionated on a 15-38% sucrose gradient layered over a 50% sucrose cushion. The polyribosome content of the 4 preparations shown are 3, 22, 55, and 54%, respectively.

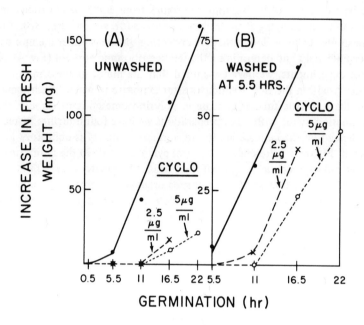

Figure 5. Effect of Cycloheximide on Embryo Germination. A) One hundred mg samples of embryo were imbibed either in water or in cycloheximide. At the indicated times, water uptake was measured as in Fig. 3A. B) One hundred mg samples of embryo were imbibed either in water or cycloheximide. At 5.5 hr, all samples were rinsed thoroughly, with vacuum suction used to dry the embryos between rinsings. All samples were returned to water. At the times indicated, samples were removed for fresh weight analysis as in A). The vacuum drying partially damages the embryos so that the water control is necessary.

What is the mechanism of the increased rate of protein synthesis? Just as with the fertilized sea urchin egg (see above) we initially considered the possibility of an impaired translational capacity. Analysis of the dry wheat embryo system with synthetic polynucleotides indicated, however, that the translational capacity of the unimbibed embryos is fully active (Marcus and Feeley, 1964). Considering the earlier observation of the marked increase in polyribosome content (Marcus and Feeley, 1965) it appeared that mRNA mobilization was the most likely explanation. Corroboration of this idea as well as the first direct demonstration of the activity of a "messenger fraction" was obtained by incubating whole homogenates of dry wheat embryos, and observing as a consequence of this incubation an "activation" of the ribosomes (Table I) (Marcus and Feeley, 1966; Marcus, 1969). The only exogenous addition required for the activation is ATP; inclusion of actinomycin D or DNase in the incubation of the homogenate has little effect on the reaction. These observations lead to the conclusion that the mobilization of mRNA is occurring, not by transcription of DNA, but rather by the activation of an mRNA already present in the wheat embryo homogenate. The study was extended to the isolation of a fraction from the homogenate that did indeed possess template activity. By a number of criteria the functional activity of the "messenger fraction" was shown to be analogous to that of an exogenous mRNA, requiring a ribosome-mRNA attachment reaction (Weeks and Marcus, 1971). More recently, we have succeeded in obtaining *in vivo* evidence for the participation of preformed mRNA in early polyribosome formation. Figure 6 shows the increase in polyribosome content, after imbibition for 30 minutes at 0° and incubation for 40 minutes at 25°. The cold pretreatment results in a decrease of the subsequent augmentation of polyribosome content but it is particularly useful in allowing the uptake of inhibitor. As detailed in a later section (see Table IV),

TABLE I

INCUBATION OF WHEAT EMBRYO HOMOGENATE
AND THE FORMATION OF ACTIVE RIBOSOMES

Viable wheat embryos were ground in a mortar in 0.02 M KCl + 0.001 M Mg-acetate. After centrifuging for 3 min at 270g, one aliquot was incubated for 12 min at 30° with ATP and an ATP-generating system while another aliquot was kept in ice. 0.5 M sucrose + 0.025 M KCl + 0.001 M Mg-acetate + 0.05 M tris (pH 7.6) + 0.005 M 2-mercaptoethanol was added to both samples and the ribosomes were isolated and assayed for incorporating activity. For complete details see Marcus and Feeley, 1966; Marcus, 1969.

Conditions	Ribosomal Activity
Complete, 30°	5140
'ATP'-omitted	285
Act D (12 μg)	5100
DNase (10 μg)	5800
RNase (0.005 μg)	2070
RNase (0.01 μg)	830
Complete, 0°	490

under these conditions of imbibition, 3'-deoxyadenosine (cordycepin) at 60 µg/ml inhibits mRNA synthesis by 60-70% while α-amanitin at 12 µM inhibits mRNA synthesis by 90%. Neither of the inhibitors shows any effect on early polyribosome formation (Fig. 6). Finally we have now determined the content of mRNA in both dry and 40-minute imbibed embryo and find them to be essentially identical. As noted in Table II, this conclusion is derived from two types of analysis. First, the fraction of embryo RNA that contains an appended poly (A) residue is determined by analysis of adsorption to a poly (dT) cellulose column (Aviv and Leder, 1972). Since almost all cellular mRNAs contain poly (A) residues (Sheldon *et al.*, 1972; Edmonds *et al.*, 1971; Lee *et al.*, 1971), this analysis in itself provides a reasonable estimate of the mRNA content. In addition there is a comparison of the template activity of total RNA from dry and imbibed embryos. These data are important since they include a measure of the activity of any mRNA that might not have an attached poly (A) residue. Again, as seen in Table II, the activities of the RNA from both the dry and the imbibed embryos are quite comparable thus establishing that the embryo mRNA content does not change significantly during early germination. Clearly polyribosome formation in the early phases of embryo germination utilizes pre-formed mRNAs.

TABLE II

mRNA CONTENT OF DRY AND IMBIBED WHEAT EMBRYOS

Embryo Sample	Yield (mg RNA/Sample)	Poly (A) + [*] (% Total RNA)	Template Activity (cpm/µg RNA)
1. 500 mg - dry	7.4	1.9	516
500 mg – 40 min 25°	6.3	2.5	500
2. 125 mg - dry	2.7		
125 mg - 40 min 25°	2.1		
3. 125 mg - dry	2.7	1.7	500
125 mg - 40 min 25°	2.4	2.0	579

[*]Fraction of total RNA adsorbed on poly (dT)-cellulose at 0.5 M NaCl and eluted with 0.01 M tris-HCl pH 7.6. The data for experiment 3 were obtained by combining the RNA samples of experiment 2 and 3 just prior to the poly (dT) cellulose analysis.

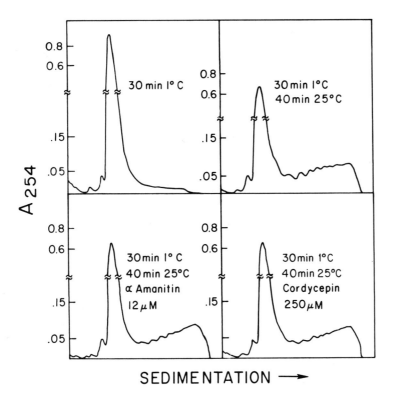

Figure 6. Effect of α-Amanitin and Cordycepin on Early Polyribosome Formation. Embryo samples were imbibed and analyzed as in Fig. 4 except that a 30-min
pretreatment at 1° was added. The polyribosome content for the 4 preparations was
6, 46, 42, and 44%, respectively.

A second plant system in which preformed mRNA appears to play an important
regulatory role is that of the cotyledons of the cottonseed. Ihle and Dure (1972a) have
shown that germination of both precocious (partially ripened) and mature (fully ripened)
cottonseed results in the *de novo* synthesis of isocitric lyase and carboxypeptidase in the
seed cotyledons. The two systems differ in that formation of the enzymes in precocious
embryos is sensitive to actinomycin D whereas enzyme formation in mature cottonseed
is insensitive to the inhibitor (Ihle and Dure, 1972b). These data have been taken to
indicate that the mRNA required for synthesis of the enzymes is transcribed at a defined
period during the ripening of the seed and that it is maintained in a non-translatable form
until the onset of germination. The system is particularly attractive for further analysis
because of the possibility of identifying the putative mRNA in terms of a specific translational product. Furthermore, indirect evidence has been obtained indicating that abscisic
acid, a plant hormone, regulates the translation of the preformed mRNA. To date, how-

ever, the mRNA has not been isolated nor is there any evidence bearing on the mechanism utilized in regulating its translation.

Function of Preformed mRNAs As Determinants of Selective Translation

Having established the reality of preformed mRNA in a number of biological systems, one is next concerned with the identification of such mRNAs in terms of their specific translational products. Thus far only a limited number of proteins have been shown to be coded for by maternal mRNAs. These include microtubule proteins (Raff *et al.,* 1971; Raff *et al.,* 1972) and histones (Gross *et al.,* 1973) in the fertilized sea urchin egg and as noted above, carboxypeptidase and isocitric lyase in the cotyledons of germinating cottonseeds. In the case of the sea urchin microtubule proteins, there is already a substantial pool in the unfertilized egg. Thus the mRNA that is utilized after fertilization or an equivalent mRNA must have been translated during oogenesis. In the case of the maternal mRNAs coding for histones, it has been shown that a large population of 9S mRNAs coding for the same proteins are made soon after fertilization (Kedes and Birnsteil, 1971). In fact, one line of evidence identifying histone mRNA in the maternal mRNA population is the ability of the RNA of the unfertilized egg to compete

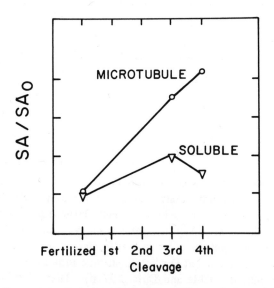

Figure 7. Changes in Relative Activity of Microtubule and Total Soluble Proteins During Early Cleavage of *Arbacia.* The specific activity of microtubule proteins, as related to a 30-min labeling interval starting at fertilization, is compared with the specific activity of total soluble protein. Actinomycin D was present at 20 μg/ml from the onset of fertilization. From R.A. Raff, H.V. Colot, S.E. Selvig, and P.R. Gross (1972).

against newly transcribed 9S mRNA in hybridization with DNA (Skoultchi and Gross, 1973). Thus both the microtubule and histone mRNAs do not function uniquely as maternal mRNAs. The other two mRNAs, *i.e.,* those proposed to function in cottonseed germination, may be more rigorous in their function. Neither carboxypeptidase nor isocitric lyase are found in the cotyledons of the ungerminated seed and, as noted above, the extent of their formation after germination seems to be unaffected by the inhibition of new transcription. Since, however, the mRNAs for these enzymes have as yet not been directly identified, their existence *per se* must be taken only as an attractive postulate.

Of further interest is the availability of the preformed mRNAs. Does the mRNA pool serve as a reservoir to be completely released when the appropriate stimulus is applied or is there a specific sequential process whereby different mRNAs are recruited into polyribosomes and as such regulate the patterns of translation? Alternatively, there might occur a gradual controlled release of mRNAs but the process would be random, with no differences in translational specificity. A complete answer to these questions is clearly not yet available. However, preliminary data as well as useful approaches have been described.

The method most useful for examining the spectrum of proteins synthesized at various times during a developmental process is polyacrylamide gel electrophoresis. Given a difference in the pattern of biosynthesis during two time periods, it is possible with this procedure to ascertain the extent to which mRNA synthesis, *i.e.* transcription, is required. To the extent that such mRNA synthesis is not required, the changing translational pattern is presumably due to recruitment of preformed mRNA. The feasibility of this approach is demonstrated in studies of both total soluble proteins (Terman, 1970) and histones (Ruderman and Gross, 1974) in fertilized sea urchin embryos, and in a study of soluble proteins synthesized during fungal spore germination (Van Etten *et al.,* 1972). With the fertilized sea urchin, it was found that a significant part of the change in protein synthetic pattern occurring between the early embryo and the blastula stages is unaffected by the presence of actinomycin D (Terman, 1970; Ruderman and Gross, 1974), thus supporting the idea of a programmed release of the preformed mRNA. A direct demonstration that the translation-level controlled changes are a consequence of mRNA mobilization should be possible by an examination of the *in vitro* products made in response to polyribosomal RNA (Ruderman *et al.,* 1974) extracted from embryos allowed to develop in the absence of transcription. Such evidence, however, has not yet been obtained.

With regard to the more limited question as to whether or not preformed mRNAs are recruited in a gradual process, a useful approach is the determination of the amount of a given mRNA that is functional at a given developmental period. A prerequisite for such an analysis is the ability to quantitate a defined translational product. The analysis could then be performed either *in vivo* with short periods of radioactive labeling or preferably in an *in vitro* incubation with mRNA isolated from the polyribosomes. A time-dependent increase in the functional content of the particular mRNA, in the absence of transcription, would support the idea of gradual recruitment. Only one such study has thus far been reported; that of the mRNA coding for microtubule protein in the fertilized sea urchin egg (Fig. 7) (Raff *et al.,* 1972). The steadily increasing specific activity of the microtubule proteins, particularly in contrast to that of the total soluble protein, suggest that this mRNA is only gradually made available to the translation system.

Mechanisms Controlling the Recruitment of Preformed mRNA

The functional significance of the preformed mRNAs requires that cell systems be able to regulate their availability to the translational system. The complexity of this "release reaction" is probably related to the mode of function of the preformed mRNAs. Thus to the extent that the access of mRNAs to the translational system is a programmed sequential process, the regulation of the release can be expected to be more complex. In approaching this question, there is one obvious starting point. Where in the cell are preformed mRNAs located and in what type of chemical structure? Attempts to answer this question have involved the fractionation of cell homogenates at neutral pH and assaying the RNA isolated from these fractions either for template activity (Gross *et al.,* 1973; Schultz *et al.,* 1972) or for the ability to hybridize to ^{3}H-poly (U) (Rosbash and Ford, 1974; Slater *et al.,* 1972; Wilt, 1973). In studies carried out with embryos of sea urchin (Gross *et al.,* 1973) and *Xenopus* (Rosbash and Ford, 1974), the mRNAs were found in cytoplasmic ribonucleoproteins sedimenting at 20S, and at 40-60S, while in similar experiments with a homogenate of dry wheat embryos (Schultz *et al.,* 1972), the mRNA activity was found in fractions sedimenting at 45S and 90S. Early after fertilization, in the case of the sea urchin, a major part of the mRNA appears to shift from the subribosomal fraction to a ribosomal location (Slater *et al.,* 1973). Two other reports examining the unimbibed wheat embryo and the unfertilized sea urchin egg systems have indicated the presence of the mRNA in a more-rapidly sedimenting fraction (Marcus, 1969; Mano and Negana, 1966). These latter studies used somewhat less labile systems of fractionation, thus allowing the maintenance of a more sensitive structural component as a possible repository of the preformed mRNA.

With regard to a specific releasing reaction, it was initially suggested (Spirin, 1966), based on the observation that the mRNAs are usually in ribonucleoprotein particles, that a proteolytic step might be involved. However, with one exception (Mano, 1966), this idea has not received further support. More recently, based on the finding that there is a doubling of the poly (A) content of sea urchin embryos after fertilization (Slater *et al.,* 1972; Wilt, 1973), it has been suggested that poly (A) synthesis is in some way involved in the activation of the preformed mRNAs. The process in the fertilized sea urchin is an attachment of poly (A) residues to maternal mRNAs in a reaction independent of transcription (Slater *et al.,* 1972; Wilt, 1973; Slater and Slater, 1974) and it occurs to a large part in the cytoplasm (Wilt, 1973; Slater *et al.,* 1973). Thus polyadenylation appears to be a reasonable candidate as a regulator of the release of preformed mRNA. Kinetic analyses, however, in which it is seen that the activation of protein synthesis precedes the polyadenylation reaction cast considerable doubt on this idea. Most recently, it has been shown (Table III), using cordycepin at a level of 750 μg/ml, that poly (A) synthesis could be completely repressed without significant effect on the activation of protein

TABLE III

**SUPPRESSION OF SYNTHESIS OF POLY (A) AND ITS EFFECT
ON PROTEIN SYNTHESIS IN SEA URCHIN EMBRYOS**[*]

Cells	Poly (A) % Total RNA	Incorporation %
Unfertilized eggs	0.033	6.8
Four-cell embryos	0.069	31.1
Embryos in cordycepin		
200 μg/ml	0.045	31.9
750 μg/ml	0.035	25.9

[*]Data from Mescher and Humphreys, 1974.

synthesis (Mescher and Humphreys, 1974). Thus at least in the case of the fertilization of sea urchin eggs, polyadenylation is not a prerequisite for the release of the preformed mRNA.

The situation with the sea urchin egg might be unique in that the change in poly (A) content after fertilization involves a change in the average size of the poly (A) tract from 100 nucleotides to 200 (Slater et $al.$, 1972). This change in size accounts for the full increment in poly (A). Conceivably, in a situation where the preformed mRNA was stored in the unadenylated form, polyadenylation could serve a regulatory function.

With the germinating seed embryo, we have also considered the possibility of polyadenylation as a regulator of the release of preformed mRNA. To test this point, we used the observation that cordycepin at 60 μg/ml and α-amanitin at 12 μM, although quite effective in inhibiting RNA synthesis (Table IV), do not affect the early increase in protein synthesis (see Fig. 6). We compared the relative incorporation of ^{3}H-uridine (U) and ^{14}C-adenosine (A) into total RNA and RNA of ribosomes in the presence and the absence of the inhibitors. If polyadenylation of preformed mRNA is a significant component of the total adenosine incorporation and if it is not inhibited by cordycepin and α-amanitin, then the ratio of A/U incorporated in the presence of the inhibitors should be

TABLE IV

COMPARATIVE INHIBITION OF URIDINE AND ADENOSINE
INCORPORATION IN TOTAL RNA AND RNA OF RIBOSOMES
BY CORDYCEPIN AND α-AMANITIN

In experiment 1, embryos were preimbibed in cordycepin $(60\,\mu g/ml)$ for 45 min at 0° and then incubated in a solution containing cordycepin $(60\,\mu g/ml)$, $[^3H]$-uridine $(150\,\mu c_i)$, and $[^{14}C]$-adenosine $(15\,\mu c_i)$ for 30 min at 0° and 8 min at 25°. Total RNA was isolated and fractionated on poly (dT)-cellulose columns into non-adsorbing (poly (A) -) and adsorbing (poly (A) +) components. In experiment 2, embryos were preimbibed in α-amanitin $(12\,\mu M)$ for 30 min at 0° and then incubated in a solution containing α-amanitin $(12\,\mu M)$, $[^3H]$-uridine $(150\,\mu c_i)$, ^{14}C-adenosine $(12\,\mu c_i)$ for 40 min at 25°. Ribosomes were isolated and the RNA was extracted from the ribosomes and fractionated as in experiment 1.

	Poly (A) -		Poly (A) +	
1. $[^3H]$-Uridine	control	cordycepin	control	cordycepin
cpm/mg RNA	1704	754	285	91
Inhibition (%)		57		68
$[^{14}C]$-Adenosine				
cpm/mg RNA	1110	562	294	125
Inhibition (%)		50		57
A/U ratio	0.66	0.75	1.07	1.37
2. $[^3H]$-Uridine	control	α-amanitin	control	α-amanitin
cpm/mg RNA	5073	1201	8954	863
Inhibition		76		90
$[^{14}C]$-Adenosine				
cpm/mg RNA	8624	2041	15223	1467
Inhibition		76		90
A/U ratio	1.7	1.7	1.7	1.7

much higher than in their absence. The data obtained (Table IV) show essentially no difference in the A/U ratio, whether or not the inhibitors are present. We therefore conclude either that there is no significant polyadenylation of preformed mRNA during the first 40 minutes of embryo germination or that the inhibitors are as effective in suppressing poly (A) synthesis as they are in preventing transcription. Since the presence of the inhibitors does not affect the activation of the preformed mRNAs (see Fig. 6) it is clear that polyadenylation is not required for this reaction. Finally, by a direct analysis of the RNA of dry and 40-minute imbibed embryos for their ability to hybridize with ^{3}H-poly (U) (Table V), we have found that in contrast to the fertilized sea urchin, there is no significant poly (A) synthesis in early embryo germination.

Are there any further suggestions for regulation of preformed mRNA release? An early observation that we made in the wheat embryo system was that the preformed mRNA could be activated by an *in vitro* incubation of a homogenate of dry embryos (see Table II), with the only exogenous addition required being ATP. We therefore considered that the level of ATP might be a regulatory component in the onset of protein synthesis during early germination. In order to test this possibility, sensitive methods were developed for the analysis of ATP, most recently a procedure using the transfer of ^{32}P-phosphoenolpyruvate to ADP. With these methods we have found that indeed there

TABLE V

POLY (A) CONTENT OF DRY AND IMBIBED WHEAT EMBRYOS

μg RNA	^{3}H-poly (U) hybridized[*] (ng)	
	dry	imbibed
6	4	4
11	13	9
28	32	24
55	45	46

[*]Data corrected for a background of 1 ng (22 cpm) hybridized in the absence of exogenous RNA.

occurs a 10-fold increase in the level of ATP during the first hour after exposure to water (Obendorf and Marcus, 1974). More recently, we have found that 90% of this increase occurs when the embryos are allowed to imbibe at 0°, a condition that does not result in polyribosome formation (see Fig. 6). Thus the increase in the ATP level, *per se,* is not sufficient to activate the translational system. In addition, analyses of the sea urchin egg system indicate that little or no change in the level of ATP occurs early after fertilization (MacKintosh and Bell, 1969). We are currently attempting to ascertain in the wheat embryo system the extent to which the increase in ATP is required for the subsequent formation of polyribosomes.

Summary

A significant feature of the early development of fertilized echinoderm and amphibian eggs and germinating seed embryos is the utilization of genetic information that has been previously transcribed during oogenesis and seed ripening. When RNA synthesis is suppressed in the early developing embryos by actinomycin D, cordycepin, or α-amanitin, there is no effect on the translation of the "preformed mRNA" thus establishing that new transcription is not necessary for early embryo development. While a wide variety of proteins appear to be coded for by "preformed mRNAs," only a limited number have been thus far identified; microtubule and histone proteins in the fertilized sea urchin egg and carboxypeptidase and isocitric lyase in germinating cottonseed. Data obtained on the protein synthetic pattern at different times after the onset of development suggest that preformed mRNAs are made available to the translational system in a gradual process, thereby providing a molecular basis for the regulation of development. The possibility is considered that polyadenylation of mRNA, a reaction known to occur early after sea urchin fertilization, is responsible for regulating the release of preformed mRNA. It is shown that this reaction (polyadenylation) can be completely suppressed with little effect on the function of preformed mRNA. Finally, it is suggested, at least for the seed embryo system, that the formation of ATP may be a prerequisite for the activation of protein synthesis.

Acknowledgement

This research was supported by U.S.P.H.S. grants GM15122, CA-06927 and RR-05539 from the National Institutes of Health; by grant GB-35585 from the National Science Foundation; and by an appropriation from the Commonwealth of Pennsylvania.

REFERENCES

Aviv, H., and Leder, P. (1972). *Proc. Natl. Acad. Sci. 69,* 1408.

Bell, E., and Reeder, R. (1967). *Biochim. Biophys. Acta 142,* 500.

Brambl, R.M., and Van Etten, J.L. (1970). *Archives Biochem. and Biophys. 137,* 442.

Ceccarini, C., Maggio, R., and Barbata, G. (1967). *Proc. Natl. Acad. Sci. 58,* 2235.

Crawford, R.B., Wilde jr., C.E., Heineman, M.H., and Hendler, F.J. (1973). *J. Embryol. Exp. Morph. 29,* 363.

Davidson, E.H., Crippa, M., Kramer, F.R., and Mirsky, A.E. (1966). *Proc. Natl. Acad. Sci. 56,* 856.

Davidson, E.H., and Hough, B.R. (1971). *J. Mol. Biol. 56,* 491.

Denny, P.C., and Tyler, A. (1964). *Biochem. Biophys. Res. Comm. 14,* 245.

Ecker, R.E., Smith, L.D., and Subtelny, S. (1968). *Science 160,* 1115.

Edmonds, M., Vaughn, M.H., and Nakazato, H. (1971). *Proc. Natl. Acad. Sci. 68,* 1336.

Epel, D. (1967). *Proc. Natl. Acad. Sci. 57,* 889.

Gambino, R., Metafora, S., Felicetti, L., and Raisman, J. (1973). *Biochim. Biophys. Acta 312,* 377.

Gross, K.W., Jacobs-Lorena, M., Baglioni, C., and Gross, P.R. (1973). *Proc. Natl. Acad. Sci. 70,* 2614.

Gross, P.R., Malkin, L.I., and Moyer, W.A. (1964). *Proc. Natl. Acad. Sci. 51,* 407.

Hultin, T. (1961). *Exptl. Cell Res. 25,* 405.

Hultin, T. (1964). *Develop. Biol. 10,* 305.

Humphreys, T. (1969). *Develop. Biol. 20,* 435.

Ihle, J.N., and Dure, L.S. (1972a). *J. Biol. Chem. 247,* 5034.

Ihle, J.N., and Dure, L.S. (1972b). *J. Biol. Chem. 247,* 5048.

Jenkins, N., Taylor, M.W., and Raff, R.A. (1973). *Proc. Natl. Acad. Sci. 70,* 3287.

Kedes, L.H., and Birnsteil, M.L. (1971). *Nature N. B. 230,* 165.

Lee, S.Y., Mendeck, J., and Brawerman, G. (1971). *Proc. Natl. Acad. Sci. 68,* 1331.

Linskens, H.F. (1969). *Planta 85,* 175.

MacKintosh, F.R., and Bell, E. (1969). *Exp. Cell Res. 57,* 71.

Mano, Y. (1966). *Biochem. Biophys. Res. Comm. 25,* 216.

Mano, Y., and Nagano, H. (1966). *Biochem. Biophys. Res. Comm. 25,* 210.

Marcus, A. (1969). *Symp. Soc. Exp. Biol. 23,* 143.

Marcus, A., and Feeley, J. (1964). *Proc. Natl. Acad. Sci. 51,* 1075.

Marcus, A., and Feeley, J. (1965). *J. Biol. Chem. 240,* 1675.

Marcus, A., and Feeley, J. (1966). *Proc. Natl. Acad. Sci. 56,* 1770.

Marcus A., Feeley, J., and Volcani, T. (1966). *Plant Physiol. 41,* 1167.

Mascarenhas, J.P., and Bell, E. (1969). *Biochim. Biophys. Acta 179,* 199.

Mescher, A., and Humphreys, T. (1974). *Nature 249,* 138.

Metafora, S., Felicetti, L., and Gambino, R. (1971). *Proc. Natl. Acad. Sci. 68,* 600.

Monroy, A., Maggio, R., and Rinaldi, A.M. (1965). *Proc. Natl. Acad. Sci. 54,* 107.

Monroy, A., and Tyler, A. (1963). *Arch. Biochem. Biophys. 103,* 431.

Nakana, E., and Monroy, A. (1958). *Exptl. Cell Res. 14,* 236.

Obendorf, R.L., and Marcus, A. (1974). *Pl. Physiol. 53,* 779.

Pollock, B.M., and Toole, V.K. (1966). *Plant Physiol. 41,* 221.

Raff, R.A., Colot, H.V., Selvig, A.E., and Gross, P.R. (1972). *Nature 235,* 211.

Raff, R.A., Greenhouse, G.A., Gross, K., and Gross, P.R. (1971). *J. Cell Biol. 50,* 516.

Rinaldi, A.M., and Monroy, A. (1969). *Develop. Biol. 19,* 73.

Rosbash, M., and Ford, P.J. (1974). *J. Mol. Biol. 85,* 87.

Ruderman, J.V., Baglioni, C., and Gross, P.R. (1974). *Nature 247,* 36.

Ruderman, J.V., and Gross, P.R. (1974). *Develop. Biol. 36,* 286.

Schultz, G.A., Chen, D., and Katchalski, E. (1972). *J. Mol. Biol. 66,* 379.

Sheldon, R., Jurale, C., and Kates, J. (1972). *Proc. Natl. Acad. Sci. 69,* 417.

Skoultchi, A., and Gross, P.R. (1973). *Proc. Natl. Acad. Sci. 70,* 2840.

Slater, D.W., Slater, I., and Gillespie, D. (1972). *Nature 240,* 333.

Slater, D.W., and Spiegelman, S. (1966). *Proc. Natl. Acad. Sci. 56,* 164.

Slater, I., Gillespie, D., and Slater, D.W. (1973). *Proc. Natl. Acad. Sci. 70*, 406.

Slater, I., and Slater, D.W. (1974). *Proc. Natl. Acad. Sci. 71*, 1103.

Smith, L.D., Ecker, R.E., and Subtelny, S. (1966). *Proc. Natl. Acad. Sci. 56*, 1724.

Spirin, A.S. (1966). *Curr. Topics in Develop. Biol. 1*, 1.

Stafford, D.W., Sofer, W.H., and Iverson, R.M. (1964). *Proc. Natl. Acad. Sci. 52*, 313.

Stavy, L., and Gross, P.R. (1967). *Proc. Natl. Acad. Sci. 57*, 735.

Terman, S.A. (1970). *Proc. Natl. Acad. Sci. 65*, 985.

Van Etten, J.L., Roker, H.R., and Davies, E. (1972). *J. Bact. 112*, 1029.

Weeks, D.P., and Marcus, A. (1971). *Biochim. Biophys. Acta 232*, 671.

Whiteley, A.H., McCarthy, B.J., and Whiteley, H.R. (1966). *Proc. Natl. Acad. Sci. 55*, 519.

Wilt, F.H. (1973). *Proc. Natl. Acad. Sci. 70*, 2345.

REGULATION OF TRANSCRIPTION IN YEAST

*C. Saunders, S.J. Sogin, D.B. Kaback and *H.O. Halvorson*

*Rosenstiel Basic Medical Sciences Research Center
and Department of Biology
Brandeis University
Waltham, Massachusetts*

Introduction

Studies on synchronous cultures of yeast have provided a large body of information in favor of periodic synthesis. This has been extensively reviewed in recent years (Donachie and Masters, 1969; Halvorson *et al.*, 1971; Mitchison, 1971; Hartwell *et al.*, 1974). Over 30-40 enzymes have been followed throughout the cell cycle in yeast. The vast majority of these show periodic synthesis restricted to some part of the cell cycle, whereas a few apparently show continuous synthesis through the cell cycle. There is no clear association of the period of enzyme synthesis with the S period. The period of synthesis is, however, spread over the cell cycle. Two major theories have been proposed to explain the control of synthesis of yeast enzymes. The "oscillatory repression" model has been proposed by several groups (Donachie and Masters, 1969; Pardee, 1966; Masters and Donachie, 1966; Goodwin, 1966) which assumes that biosynthetic enzymes are controlled by end-product repression. Given appropriate constants, a negative feedback system will oscillate. The supporters of this hypothesis have argued that the oscillations are entrained by a cell cycle dependent event. This theory fits much of the data in synchronous bacterial cultures (Donachie and Masters, 1969; Halvorson *et al.*, 1971; Mitchison, 1971) and some cases in eukaryotic cells such as the ribulose 1,5-diphosphate carboxylase in *Chlorella* (Malloy and Schmidt, 1970). However, oscillatory repression does not fit several enzymes in yeast. Ornithine transcarbamylase is a step enzyme in fission yeast, however, Stebbing (1972) found no fluctuations in the amino acid pools during the cell cycle of either repressed of constitutive synthesis. Similarly, arginase and ornithine transaminase in *Saccharomyces cerevisiae* shows steps at the same stage of the cycle both with induction by arginine and without, but no major cyclic changes in the argnine pool were observed (Carter *et al.*, 1971).

---

*Symposium participant.

We have developed a second theory of control of periodic synthesis of enzyme which assumes an ordered transcription of the genome (Halvorson, 1966). We found that the time in the cell cycle at which a step occurs was not influenced by increasing gene dosage at one locus. When non-allelic genes for the same enzyme were introduced, extra steps were observed (Tauro and Halvorson, 1966). The correlation between the step timings and gene positions on chromosome V suggested that the genes were transcribed in a linear or sequential manner (Tauro *et al.*, 1968). This was supported by the findings of Cox and Gilbert (1970) in which the distance between two enzyme genes on the second chromosome is much greater in one strain than in the other and that there is a corresponding change in the distance between the two steps in the cell cycle. Similar evidence favoring linear readings have been observed in synchronous spore germination with 5 enzymes in *Bacillus subtilis* (Kennett and Sueoka, 1971). In contrast, as will be discussed below, in *Saccharomyces cerevisiae* step functions are not observed for ribosome synthesis. The synthesis of ribosomal RNA (Williamson and Scopes, 1960; Halvorson *et al.*, 1964) and ribosomal proteins (Shulman *et al.*, 1973) is continuous during the yeast cell cycle. Recently, we also observed that polymerase I, the enzyme responsible for ribosomal RNA transcription, is continuous during the cell cycle (Sebastian *et al.*, 1974), while polymerase II was a step enzyme.

The intriguing question is why so many of the enzymes in *Saccharomyces cerevisiae* follow a periodic step increase. Temporal expression has also been implicated to explain the morphogenic changes during the cell cycle and the behavior of cell cycle mutants (Hereford and Hartwell, 1974). There has been considerable discussion on the nature of a cellular clock whose timing is essential for progress through the cell cycle (Donachie and Masters, 1969; Halvorson *et al.*, 1971; Mitchison, 1971). We have, however, little basic information at the molecular level concerning the mechanism. For example, is the periodic synthesis of an enzyme a consequence of the discontinous transcription of its gene?

mRNA Synthesis and Stability During the Cell Cycle

The half-life of mRNA of yeast has only been determined indirectly. Employing a temperature sensitive mutant (ts 136) in which little RNA synthesis occurs at the nonpermissive temperature, Hartwell *et al.* (1970) estimated a half-life of 23 minutes for yeast mRNA from the kinetics of polyribosome disappearance after transferring cells to the restrictive temperature. Similar conclusions were reached following the decay in protein synthetic ability (Tonnesen and Friesen, 1973; Cannon *et al.*, 1973) or specific enzyme synthesis (Kuo *et al.*, 1973) following addition of inhibitors of RNA synthesis. However, Lawther and Cooper (1973) reported ca. a three minute half-life in the capacity to synthesize allophanate hydrolase in yeast suggesting the possibility of mRNAs with different stabilities. There is little direct information on mRNA synthesis or stability during the cell cycle in yeast, in contrast to work performed in other developing systems. In bacteriophage T_4 and λ, direct evidence is available that mRNA populations vary during their cell cycle. Although we have observed fluctuations in base composition of rapidly labeled RNA during the yeast cell cycle (unpublished results), it is not possible to determine the complexity or stability of mRNA at various stages. Alternatively, molecular hybridization has been used to identify mRNA molecules during the cell cycle. This was first employed by Cutler and Evans (1966) who pulse labeled synchronous cultures of *E. coli* with BUdR to create DNA components of high density each representing different segments of the chromosome. These were isolated and used as DNA inputs for filter hybridization. Their results suggested that specific mRNAs were synthesized at various stages of the cell cycle. Hansen *et al.* (1970) derived similar conclusions using synchronous cultures obtained from outgrowing spores of *B. cereus*.

Both of these techniques suffer from the need to consider large areas of the genome with the result that a heterogenous messenger population was analyzed. The lack of precision in these results is disappointing since the concept of sequential transcription has been advanced in bacteria (Kennett and Sueoka, 1971) as well as in yeast (Tauro *et al.*, 1968). These findings suggest that specific messengers are synthesized during a restricted portion of the cell cycle. While the findings previously cited support this interpretation on an overall basis, quantitation of this phenomenon as well as restriction of this view to a single messenger, has yet to be established.

The majority of information regarding the synthesis and degradation of specific mRNAs in eukaryotic cells has been derived from systems involving RNA viruses. In all systems studied to date, these messengers have quite long half-lives (on the order of hours) as compared to prokaryotic cells (on the order of minutes). These findings have indicated that the regulation of protein synthesis may be quite different in eukaryotic cells. Perlman *et al.* (1972) have suggested that the regulation of protein synthesis in the adeno-virus system is a translational control as opposed to a transcriptional control characteristic of prokaryotes.

It would be of interest to analyze messenger material made during specific portions of the cell cycle in yeast. Yeast can be easily manipulated to yield large numbers of cells at specific stages in the cell cycle. These techniques have been used to analyze the rate of synthesis of rRNA during the cell cycle (Sogin *et al.*, 1974). In addition, by simple nutritional deprivation yeast can undergo a meiotic developmental cycle resulting in the formation of ascospores. During this developmental cycle, the rate of processing of ribosomal RNA slows from a completion time of twenty minutes (Udem and Warner, 1972) to over 150 minutes (Sogin *et al.*, 1972). It is also of interest to note that six hours after the cells were placed in sporulation medium, a class of mRNA was synthesized at a rate exceeding that of rRNA. This RNA was unstable and was degraded during the subsequent chase (Sogin *et al.*, 1972). This result indicated that the pattern of RNA synthesis was markedly different during meiosis compared to the pattern of RNA synthesis during the vegetative cycle. Although the 20S RNA synthesized during development was originally thought to be a stable messenger, subsequent analyses have shown this RNA to be a submethylated precursor to the 18S RNA. The existence of this rRNA species is an indication that at least the rRNA synthesis is aberrant during development.

Regulation by Poly(A)

Polyadenylation is a feature of some but not all of the messenger RNAs from a variety of eukaryotic organisms including yeast (Brawerman, 1974; McLaughlin *et al.*, 1973). While the function of these sequences remains unknown, there is some evidence which suggests that they may play a role in messenger transport out of the nucleus and in regulating the half-life of existing cytoplasmic messengers. Darnell *et al.* (1971) have reported an enrichment of poly(A)-containing species in the cytoplasm over HnRNA. The suggestion here, of course, is that the poly(A) region plays a role in the selection of messenger for transport. This is consistent with the observation of Blobel (1972) that a 73,000 dalton protein is found associated with the poly(A) region of messenger transported to the cytoplasm. Experiments utilizing 3'-deoxyadenosine (cordycepin) indicate that inhibition of the nuclear poly(A) polymerase also blocks a large fraction of messenger transport (Weinberg, 1973) as well as enzyme induction (Sarkar *et al.*, 1973). Histone mRNA is undoubtedly the best characterized non-poly(A)-containing messenger. The kinetics of appearance of this species in the cytoplasm does not show the lag time of transport which the poly(A)-containing messengers do. This is presumably a reflection of the time required for the post-transcriptional polymerization of poly(A) to the 3' end of these messages.

The first evidence for poly(A) regulation of messenger half-life comes from the observation that de-adenylated globin messenger RNA injected into *Xenopus* oocytes begins to lose template activity after 1 hour while the poly(A)-containing message persisted at the initial velocity for 48 hours (Huez *et al.*, 1974). It is important to note, however, that the initial velocity was identical in both cases. This is consistent with the data obtained with de-adenylated message in *in vitro* translational systems (Munoz and Darnell, 1974; Bard *et al.*, 1974). Earlier work in this field, while indicating that histone messenger had a shorter half-life than the bulk poly(A)-containing messenger must be subject to possible re-interpretation owing to possible actinomycin D artifacts (Singer and Penman, 1972).

The gradual shortening of poly(A) lengths on messenger observed *in vivo* may then be a mechanism for disposing of messengers no longer necessary for cellular functions. A poly(A) ase activity has been observed associated with polysomes from calf thymus (Rosenfeld *et al.*, 1972). The cytoplasmic form of poly(A) polymerase would, of course, work to balance the degradation of poly(A). This leads to an attractive control mechanism with offsetting synthesis and degradation necessary to maintain a critical poly(A) length required for continued message function.

RNA Synthesis During the Cell Cycle

From the early experiments in synchronously dividing yeast by Williamson and Scopes (1960), and later by Halvorson *et al.* (1964), it is clear that total RNA, ribosomal RNA, tRNA and total protein increase continuously during the cell cycle. Because of the fundamental differences between the kinetics of RNA synthesis and the pattern of step enzymes during the cell cycle, a re-examination of the kinetics of the transcription processes was warranted. This re-examination was possible since one can now directly examine cell cycle dependent processes in random exponential cultures thereby avoiding possible physiological alterations arising from methods used to obtain synchronous division. This is achieved by separation of an exponential culture on a sucrose gradient in a zonal rotor into size classes representing cell cycle stages (Sebastian *et al.*, 1971). It was of interest to determine whether the amount of rRNA increases continuously or stepwise during the cell cycle. A culture pregrown on $[^{14}C]$ adenine was pulsed for 10 minutes with $^{32}PO_4$, the rRNA isolated and fractionated on 3% polyacrylamide gels. In analyzing the gels the $[^{14}C]$ adenine label of the 18S RNA measures the amount of this RNA present in the extract from cells at a particular stage in the cell cycle, while the ^{32}P label estimated the amount of 18S RNA made and processed in a 10 minute pulse. Thus, the $^{32}P/^{14}C$ ratio of the 18S RNA is a measure of the rate of RNA synthesis at different stages of the cell cycle. The $^{32}P/^{14}C$ ratio remained constant throughout the cell cycle demonstrating that rRNA is not only made in all stages of the cell cycle but at an increasing rate throughout the cell cycle (confirming earlier results).

One possible interpretation of these results is that the increased rate of rRNA synthesis observed is a result of either changes in the level of precursor pools or of isotope uptake leading to changes in the specific activities of the RNA precursors. These possibilities were eliminated by examining the nucleotide pools during the cell cycle (Sogin *et al.*, 1974). The nucleotide pools were labelled for two generations with carrier-free ^{33}P. The culture was then exposed to carrier-free ^{32}P for 10 minutes and separated according to stages in the cell cycle by zonal centrifugation. Nucleotide pools were extracted and fractionated and the $^{33}P/^{32}P$ ratio of the nucleoside triphosphates determined. Although there is some fluctuation in individual nucleosides, the specific activities of the total nucleoside triphosphate pools remain approximately constant throughout the cell cycle.

Some clues on control of rRNA synthesis could be gained by quantitative considerations. Ribosomal DNA in yeast contains 140 reiterated genes for rRNA (Finkelstein *et al.*, 1972; Schweizer *et al.*, 1969; Øyen, 1973). From measurements of the rate of precursor synthesis of RNA synthesized per generation (Udem and Warner, 1972; Mortimer and Hawthorne, 1973) and from the number of cistrons for rRNA (rDNA) one can readily calculate that all or most of the rDNA genes are transcribed during the cell cycle. To examine the regulation of rRNA synthesis, we have (a) studied the location and distribution of rRNA genes and (b) studied the level, activity, and properties of yeast RNA polymerases.

rDNA

Two components of nuclear DNA have been distinguished by their different densities on CsCl gradients of DNA from *Saccharomyces cerevisiae*. The major component

(αDNA) has a buoyant density of 1.699 g/cm^3 and the dense satellite (γDNA) has a density of 1.705 g/cm^3 (Cramer et al., 1972). The γDNA represents about 12% of the total nuclear DNA and contains most if not all of the ribosomal RNA genes. Retel and Planta (1972) and Ruben and Sulsten (1973) have reported that γDNA also contains the 5S RNA cistrons and a small proportion of the tRNA cistrons. α and γDNA can be separated and purified by centrifugation on Hg^{2+}Cs$_2$SO$_4$ density gradients or Pt(NH$_3$)$_2$Cl$_2$/ CsCl (Stone et al., 1975) and the complementary strands of DNA separated in alkaline CsCl (Cramer et al., 1972). Cramer et al. (1972) have recently demonstrated that the information coding for the ribosomal RNA is located in the light (L) strand of the γDNA.

To locate rRNA cistrons, DNA-RNA hybridization experiments were carried out with DNA isolated from disomic (n+1) and monosomic ($2n$-1) strains. If the dosage of a chromosome carrying many rRNA genes is altered, one would expect an effect on the percent DNA hybridization by rRNA. Experiments with a number of disomic strains of yeast failed to locate these clusters on any of a number of chromosomes (Goldberg et al., 1972; Gimmler and Schweizer, 1972). Øyen initially observed that DNA extracted from a strain monosomic ($2n$-1) for chromosome I had less rDNA than that present in a related diploid ($2n$) strain (Goldberg et al., 1972). Subsequently, rRNA-DNA hybridization experiments in three different laboratories have associated 60-70% of the rDNA with chromosome I (Finkelstein et al., 1972; Schweizer et al., 1969; Kaback et al., 1973). By DNA-RNA hybridization and equilibrium density centrifugation of sheared nuclear DNA, Cramer et al. (1972) concluded that the rDNA genes are present in clusters of between 10-30 cistrons. Assuming 140 rDNA genes per haploid genome, it was concluded that there are between 80-100 rDNA genes on chromosome I in 3-10 clusters spaced by DNA with a higher (A + T)/(G + C) base content than rDNA. Recently the size of chromosome I has been estimated to be 400-450 x 10^6 daltons (Finkelstein et al., 1972). This number of rDNA genes on chromosome I would therefore approximate the size of this chromosome; however, the accuracy of these experiments do not preclude a significant amount of chromosome I coding for other functions, including the region spacing of rRNA genes.

As shown by genetic studies, chromosome I is the smallest yeast chromosome containing 3 centromere linked markers no further than 5-10 centimorgans distal to the centromere. Since the rRNA genes account for DNA of chromosomal size and since these genes appear to be located in clusters on chromosome I, one might expect other genetic markers scattered on the chromosome. Since rRNA genes are 140-fold reiterated and contain no known heterogeneity, any lesion in a single rRNA gene should not be expressed as a mutation.

To carry out the saturation mapping of chromosome I, we have developed a method which enriches for mutations on this chromosome. The method is to mutagenize a strain monosomic ($2n$-1) for chromosome I. Only mutations arising on chromosome I and dominant mutations on other chromosomes would show a mutant phenotype. A number of such temperature sensitive (ts) mutations have been isolated. In an initial genetic study of five of these, two fell into the same complementation group. None of the five mutants were linked to ade_1, the chromosome I centromere marker. However, we were able to show that the two non-complementing mutants were on chromosome I by mitotic recombination and disome exclusion mapping (Mortimer and Hawthorne, 1973).

Strains heterozygous for ade_1 and the thermosensitive mutants were constructed by crossing strains that were ade_1, and ts to appropriately marked haploids which were ADE. Mitotic recombination was induced by brief treatment with ultraviolet light. Colonies were grown up and red sectors homozygous for ade_1, due to a mitotic crossover, were picked. If the ts allele is located on the same arm as the ade_1 allele then it too would be homozygous due to the mitotic crossover and the red sector would be thermosensitive. Co-sectoring of the two phenotypes ade and thermobility occurred for two of the five complementation groups isolated, evidence that these two markers were on chromosome I.

We further confirmed the above evidence by looking at the segregation of trisome $(2n + 1)$-disome exclusion for chromosome I. The mutants gave the expected segregation when a ts/+/+ trisome was dissected: 4:0, 3:1 and 2:2, and WT:TS asci. Therefore, the mutants gave us a new "genetic fragment" which is located equal or greater than 50 centimorgans distal to the centromere and ade_1 on chromosome I. Conditional mutants are readily isolated by this procedure. To date over 160 ts mutants have been isolated.

RNA Polymerases in Yeast and Their Regulation

The presence of multiple RNA polymerases in eukaryotic organisms, including yeast (Ponta et al., 1971; Brogt and Planta, 1972; Dezelee et al., 1972; Ademetal, 1972; Sebastian et al., 1973) suggests that their functions in transcribing the genome may be different. Mammalian and amphibian RNA polymerase I (resistant to α-amanitin) is located in the nucleolus (Roeder and Rutter, 1970; Roeder et al., 1970) and this is the site of synthesis and processing of rRNA genes (Ritossa and Spiegelman, 1965). These observations have led to the suggestion that RNA polymerase I is involved in the transcription of ribosomal DNA *in vivo*.

Purified preparations of yeast nuclei yielded three polymerase species of DNA-dependent RNA polymerase. Enzymes I, II, and III can be separated by DEAE-Sephadex chromatography. Using a stepwise $(NH_4)_2SO_4$ gradient, enzymes I, II and III are eluted respectively at 0.2, 0.25 and 0.35 M $(NH_4)_2SO_4$. Earlier reports describing detection of only two enzymes, I and II (Sebastian et al., 1973) can be attributed to the use of NH_4Cl during purification on DEAE-Sephadex. It has been shown by us (unpublished observations) that enzyme II, eluted on DEAE-Sephadex with NH_4Cl, can be subsequently separated into enzymes II and III on rerunning on DEAE-Sephadex using $(NH_4)_2SO_4$. RNA polymerase I transcribed the L strand of the ribosomal DNA preferentially *in vitro* when either γDNA or whole nuclear DNA was used as a template, whereas RNA polymerase II transcribed both strands equally (Cramer et al., 1974). The base composition of the polymerase I product resembles that of ribosomal RNA and the size distribution of

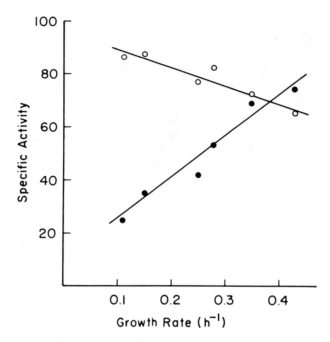

Figure 1. Specific activity of RNA polymerase I (●) and II (○) in *S. cerevisiae* IL46 grown at different growth rates in the chemostat. Data from Sebastian *et al.* (1973).

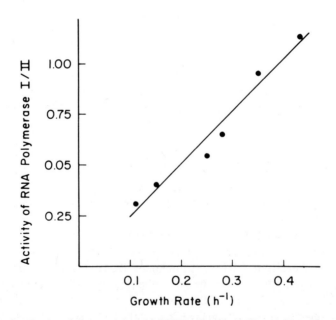

Figure 2. Ratio of the activity of RNA polymerase I and II in *S. cerevisiae* IL46 grown at different growth rates. Data from Sebastian *et al.* (1973).

the largest species (30-35S) is roughly equivalent to the size of the largest ribosomal RNA precursor synthesized in vivo (Udam and Warner, 1972). In yeast there is a direct relationship between the polymerase I level and the growth rate. Over the range studied, Sebastian et al. (1973) investigated the relationship of rRNA synthesis to the growth rate and RNA content per cell. Since 85% of the total yeast RNA is ribosomal RNA (Warner, 1971), the variations in the cellular RNA content observed are mainly due to variations in the amount of rRNA. The control mechanisms involved in the regulation of RNA synthesis at different growth rates are as yet unknown. The possibility was eliminated that different rates of synthesis of rRNA at various growth rates was due to differential amplifications of the ribosomal RNA cistrons (Schweizer and Halvorson, 1969). However, there appears to be a direct relationship between the polymerase I level and the growth rate (Fig. 1, Fig. 2). As determined in a chemostat, slow growing cells contain proportionally less RNA polymerase I than fast growing cells while there were only slight changes in the level of RNA polymerase II with changes in the growth rate. No differences were found in the template specificity, metal requirements, salt optimum and α-amanitin sensitivity of RNA polymerase I and II from either slow or fast growing cells. This relationship suggested that the synthesis of rRNA, the major component of the cellular RNA, can be regulated in fast and slow growing yeast cells by the level of RNA polymerase I activity.

We have recently found (Fig. 3) that the ratio of the activities of the nuclear RNA polymerases I and II are different at different stages of the cell cycle indicating that these enzymes are regulated independently (Sebastian et al., 1974). The specific activity of the RNA polymerase I remained constant during the cell cycle suggesting that its activity in culture increased continuously and that the synthesis of the enzyme or some factor required for its activity was continuous. On the other hand, the changes in specific activity of the RNA polymerase II indicated a discontinuous synthesis of this enzyme (or factors) during the cell cycle. The synthesis of rRNA (Williamson and Scopes, 1960; Halvorson et al., 1964) and of the ribosomal proteins (Shulman et al., 1973) is continuous during the yeast cell cycle, while replication of the DNA as well as the rDNA (Gimmler and Schweizer, Unpublished results) and the synthesis of many enzymes occur in a discontinuous, stepwise fashion (Halvorson et al., 1971; Mitchison, 1971).

Thus, most of the rRNA genes are transcribed continuously during the cell cycle and the rate of rRNA synthesis is not immediately affected by gene dosage.

The fact that there is no detectable gene dosage effect on the rate of ribosomal RNA synthesis in yeast indicates that the increasing rate throughout the cell cycle is controlled by other factors. One possibility is that the entire chromosome I is continuously accessible to transcription and that ribosomal RNA polymerase increases continuously throughout the cell cycle resulting in increasing initiation of transcription/gene/minute. In other eukaryotic systems, polymerases I and II are independently regulated by hormones (Smuckler and Jacob, 1971; Sajdel and Jacob, 1971). In yeast it has been suggested that RNA polymerase I levels might be proportional to the level of RNA (Gimmler and Schweizer, Unpublished results). However, other possibilities include increases during the cell cycle of regulators of RNA synthesis such as psi factors or guanosine tetraphosphate (Enger and Tobey, 1969; Gallant *et al.,* 1971) or specific repressor molecules. Although we do not have an understanding of the mechanism of regulation, recent observations in our laboratory (Sogin *et al.,* 1974) indicate the existence of two molecular species of 26S RNA in yeast. Sogin *et al.* (1974) observed that the "smaller" of the two 26S RNA species is preferentially made later in the cell cycle. Sugiura and Takanami (1967) suggested on the basis of nucleotide sequencing in *Saccharomyces cerevisiae* that there were two distinct types of 26S ribosomal RNA, one having 5′-nucleotide sequence pApApApCp and the other PuPxpXpXpXpGp while there was only one molecular species of 18S RNA. The interesting possibility therefore exists that yeast may possess more than one species of 26S ribosomal RNA cistron which may be differentially transcribed or processed during the cell cycle. Although we have little information regarding this

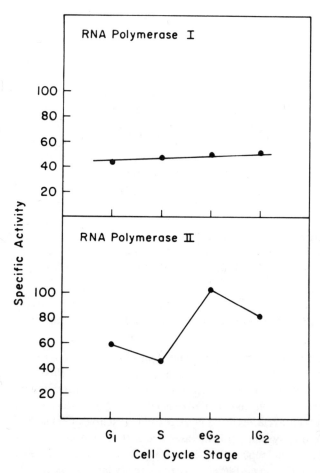

Figure 3.   Specific activity of the RNA polymerase I and II from cells at different stages of the cell cycle of *S. cerevisiae* Y185. Data from Sebastian *et al.* (1974).

possibility, evidence was discussed above for more than one chromosome localization of ribosomal RNA cistrons in yeast.

Conclusions

1. The majority of the genes for ribosomal RNA are concentrated in clusters on chromosome I.

2. Both rRNA and polymerase I are synthesized continuously over the cell cycle and the level of RNA polymerase I is proportional to the rate of rRNA synthesis.

3. The regulation of rRNA and mRNA synthesis differ.

Acknowledgements

This investigation was supported by a USPHS Grant AI 01459 from the National Institute of Allergy and Infectious Diseases (H.O. Halvorson).

REFERENCES

Adam, R., Schultz, L.D. and Hall, B.D. (1972). *Proc. Natl. Acad. Sci.* USA *69*, 1702.

Bard, E., Efron, D., Marcus, A. and Perry, R.P. (1974). *Cell 1*, 101.

Blobel, A. (1972). *Biochem. Biophys. Res. Comm. 47*, 88.

Brawerman, A. (1974). *Annu. Rev. Biochem. 43*, 621.

Brogt, T.M. and Planta, R.J. (1972). *FEBS Letters 20*, 47.

Cannon, M., Davies, J.E. and Jimenez, A. (1973). *FEBS Letters 32*, 277.

Carter, B.L.A., Sebastian, J. and Halvorson, H.O. (1971). *Advances in Enzyme Regulation 9*, 253.

Cox, C.G. and Gilbert, J.B. (1970). *Biochem. Biophys. Res. Comm. 38*, 750.

Cramer, J.H., Bhargava, M.M. and Halvorson, H.O. (1972). *Anal. Biochem. 49*, 276.

Cramer, J.H., Bhargava, M.M. and Halvorson, H.O. (1972). *J. Mol. Biol. 71*, 11.

Cramer, J.H., Sebastian, J., Rownd, R.H. and Halvorson, H.O. (1974). *Proc. Natl. Acad. Sci.* USA *71*, 2188.

Cutler, R.G. and Evans, J.E. (1966). *J. Mol. Biol. 26*, 91.

Darnell, J., Wall, R. and Tushinski, R. (1971). *Proc. Natl. Acad. Sci. 68,* 1321.

Dezelee, S., Sentenac, A. and Formageot, P. (1972). *FEBS Letters 21,* 1.

Donachie, W.D. and Masters, M. (1969) in *The Cell Cycle, Gene-Enzyme Interactions,* eds. Papilla, G.M., Whitson, G.L. and Cameron, I.L. (New York and London: Academic Press), p. 37.

Enger, M.D. and Tobey, R.A. (1969). *J. Cell Biol. 42,* 308.

Finkelstein, D.B., Blamire, J. and Marmur, J. (1972). *Nature New Biology 240,* 279.

Gallant, J., Irr, J. and Castel, M. (1971). *J. Biol. Chem. 246,* 5812.

Gimmler, G.M. and Schweizer, E. (1972). *J. Mol. Biol. 72,* 811.

Goldberg, S., Øyen, T., Idriss, J.M. and Halvorson, H.O. (1972). *Mol. Gen. Genet. 116,* 139.

Goodwin, B.C. (1966). *Nature 209,* 479.

Halvorson, H.O., Bock, R.M., Tauro, P., Epstein, R. and LaBerge, M. (1966) in *Cell Synchrony,* eds. Cameron, I.L. and Padilla, G.M. (New York and London: Academic Press), p. 102.

Halvorson, H.O., Carter, B.L.A. and Tauro, P. (1971). *Advances in Microbial Physiology 6,* 47.

Halvorson, H.O., Gorman, J., Tauro, P., Epstein, R. and Berge M. (1964). *Fed. Proc. 23,* 1002.

Hansen, J.N., Spiegelman, G. and Halvorson, H.O. (1970). *Science 168,* 1291.

Hartwell, L.H., Culotti, J., Pringle, J.R. and Reid, B.J. (1974). *Science 183,* 46.

Hartwell, L.H., Hutchison, H.T., Holland, T.M. and McLaughlin, C.S. (1970). *Molec. Gen. Genet. 106,* 347.

Hereford, L. and Hartwell, L.H. (1974). *J. Mol. Biol. 84,* 445.

Huez, G., Marbaix, G., Hubert, E., Leclercq, M., Nudel, U., Soreq, H., Salomon, R., Lebleu, B., Revel, M. and Littauer, U.Z. (1974). *Proc. Natl. Acad. Sci. USA, 71,* 3143.

Kaback, D.B., Bhargava, M.M. and Halvorson, H.O. (1973). *J. Mol. Biol. 79,* 735.

Kennett, R.H. and Sueoka, N. (1971). *J. Mol. Biol. 60,* 31.

Kuo, S.C., Cano, F.R. and Lampen, J.O. (1973). *Antimicrob. Ag. Chemother. 3,* 716.

Lawther, R.P. and Cooper, T.G. (1973). *Biochem. Biophys. Res. Comm. 55,* 1100.

Masters, M. and Donachie, W.D. (1966). *Nature 209,* 476.

McLaughlin, C.S., Warner, J.R., Edmonds, M., Nakazato, H. and Vaughn, M.H. (1973). *J. Biol. Chem. 248,* 1466.

Mitchison, J.M. (1971). *The Biology of the Cell Cycle* (London: Cambridge Univ. Press.)

Molloy, G.R. and Schmidt, R.R. (1970). *Biochem. Biophys. Comm. 40,* 1125.

Mortimer, R.K. and Hawthorne, D.C. (1973). *Genetics 74,* 33.

Munoz, R.E. and Darnell, J.E. (1974). *Cell 2,* 247.

Øyen, T.B. (1973). *FEBS Letters 30,* 53.

Pardee, A.B. (1966). *Metabolic Control Colloquium of the Johnson Research Foundation* (New York: Academic Press), p. 239.

Perlman, S., Hirsch, M. and Penman, S. (1972). *Nature New Biology 238,* 143.

Ponta, H., Ponta, U. and Wintersberger, E. (1971). *FEBS Letters 18,* 204.

Retel, J. and Planta, R.J. (1972). *Biochim. Biophys. Acta. 281,* 299.

Ritossa, F.M. and Spiegelman, S. (1965). *J. Mol. Biol. 12,* 829.

Roeder, R.G., Reeder, R.H. and Brown, D.D. (1970). *Cold Spring Harbor Symp. Quant. Biol. 35,* 727.

Roeder, R.G. and Rutter, W.J. (1970). *Proc. Natl. Acad. Sci.* USA *69,* 1702.

Rosenfeld, M.G., Abrass, I.B. and Perkins, L.A. (1972). *Biochem. Biophys. Res. Comm. 49,* 230.

Rubin, G.M. and Sulstan, J.E. (1973). *J. Mol. Biol. 79,* 251.

Sajdel, E.M. and Jacob, S.T. (1971). *Biochem. Biophys. Res. Comm. 45,* 707.

Sarkar, P.K., Goldman, B., Moscona, A.A. (1973). *Biochem. Biophys. Res. Comm. 50,* 308.

Schweizer, E. and Halvorson, H.O. (1969). *Exptl. Cell Res. 56,* 239.

Schweizer, E., MacKenchnie, C. and Halvorson, H.O. (1969). *J. Mol. Biol. 40,* 261.

Sebastian, J., Bhargava, M.M. and Halvorson, H.O. (1973). *J. Bact. 114,* 1.

Sebastian, J., Carter, B.L.A. and Halvorson, H.O. (1971). *J. Bact. 108,* 1045.

Sebastian, J., Mian, F. and Halvorson, H.O. (1973). *FEBS Letters 34,* 159.

Sebastian, J., Takano, I. and Halvorson, H.O. (1974). *Proc. Natl. Acad. Sci.* USA *71,* 769.

Shulman, R.W., Hartwell, L.H. and Warner, J.R. (1973). *J. Mol. Biol. 73,* 513.

Singer, R.H. and Penman, S. (1972). *Nature 240,* 100.

Smuckler, E.A. and Jacob, S.T. (1971). *Nature 234,* 37.

Sogin, S.J., Carter, B.L.A. and Halvorson, H.O. (1974). *Exptl. Cell Res. 89,* 127-138.

Sogin, S.J., Haber, J.E. and Halvorson, H.O. (1972). *J. Bact. 112,* 806.

Stebbing, N. (1972). *Exptl. Cell Res. 70,* 381.

Stone, P.J., Sincer, F.M., Kelman, A.D., Bhargava, M.M. and Halvorson, H.O. (1975). To be submitted for publication.

Sugiura, M. and Takanami, M. (1967). *Proc. Natl. Acad. Sci.* USA *58,* 1595.

Tauro, P. and Halvorson, H.O. (1966). *J. Bact. 92,* 652.

Tauro, P., Halvorson, H.O. and Epstein, R.L. (1968). *Proc. Natl. Acad. Sci.* USA *59,* 277.

Tonnesen, T. and Friesen, J.D. (1973). *J. Bact. 115,* 889.

Udem, S. and Warner, J. (1972). *J. Mol. Biol. 65,* 227.

Warner, J.R. (1971). *J. Biol. Chem. 246,* 447.

Weinberg, R.A. (1973). *Annu. Rev. Biochem. 42,* 329.

Williamson, D.H. and Scopes, A.W. (1960). *Exptl. Cell Res. 20,* 338.

NUCLEAR TRANSPLANTATION AND THE ANALYSIS OF GENE ACTIVITY IN EARLY AMPHIBIAN DEVELOPMENT

J.B. Gurdon

M.R.C. Laboratory
Hills Road, Cambridge, England

Amphibian eggs are large and tolerant to the injection of small pipettes. For this reason they are particularly well suited for microinjection experiments. This paper discusses certain types of injection experiments that have proved informative in the analysis of gene activity in development.

The three major levels at which gene expression can be regulated in development are (1) replication, (2) transcription, and (3) translation. Regulation at the level of replication could result in genes being unequally distributed to different cell-types. Regulation of transcription would lead to the production of different messenger RNAs in different cells which contain the same genes. Regulation at the level of translation could cause different proteins to be synthesized from the same population of messenger RNAs. One of the first problems in development is to find out which, if not all, of these levels is used to regulate gene expression in development. The transplantation of nuclei to enucleated eggs is a type of manipulation which has provided strong evidence for the identity of the genome in specialised cells. The first part of this article summarises some recent experiments which strongly reinforce the evidence for this point of view. Transcriptional control is only now beginning to be approachable by the use of microinjection methods. The direction of such attempts are discussed in the second half of this article. The injection of messenger RNAs into eggs and oocytes has permitted some clear cut experimental tests of the importance of translational control in development. Message injection experiments in oocytes have recently been summarized by Gurdon (1974a). The experimental introduction of messenger RNA into fertilized eggs, and hence into specialised cells, has been described by Gurdon *et al.* (1974) and Woodland *et al.* (1974).

Methods of Nuclear Transplantation

The technique of nuclear transplantation has been operated successfully, within the vertebrates, in only certain species of amphibia. The experiments to be summarized here were carried out with *Xenopus,* in which genus the technique of nuclear transplantation was originally described by Elsdale, Gurdon and Fischberg (1960). A few modifications to the technique have been introduced in recent years, and these are specified in the papers to be referred to.

Donor Cells for Nuclear Transplantation

The first kind of fully specialised cell-type from which nuclei could be successfully transplanted was an intestinal epithelium cell of feeding *Xenopus* larvae (Gurdon, 1962). The most recent work on *Xenopus* has involved the use of nuclei from adult skin cells. These have been chosen because they can be obtained as a very homogeneous population in which virtually every cell can be demonstrated to contain keratin at the time of nuclear transplantation. As far as the future analysis of nuclear-transplant embryos is concerned (see below), skin cells have a major advantage over intestinal epithelium cells; they synthesize a characteristic and well-defined protein (keratin) which can be assayed sensitively. To obtain a homogeneous population of keratin-synthesizing skin cells, cultured explants must be used. When a small piece of foot-web skin from *Xenopus* is placed in culture, cells migrate out from it to form a monolayer sheet within a few days. Reeves and Laskey (1975) have shown that if the outgrowth takes place under standard conditions, some of the cells which migrate out divide and fail to undergo keratinization at the same time as most of these cells do. Eventually all of these, evidently epidermal, cells keratinize, and we therefore describe them as "determined" since they are committed to undergo skin differentiation. If the conditions of outgrowth are altered, most significantly by omission of plasma from the medium, the outgrowth monolayer of cells undergoes synchronous keratinization, such that virtually all of the outgrowth cells contain keratin on the third day after the explant was made. Cells of this type are described as 'differentiated' skin cells since they contain a type of protein peculiar to specialised epidermal cells.

The criterion by which skin cell differentiation is defined is very important. Reeves (1975) has purified a protein of about 69,000 molecular weight from adult frog skin, and has prepared antibodies against it. The antibody preparation was made fluorescent and tested in a variety of differentiated cell-types. Reeves (1975) found that the antikeratin antibody bound only to skin cells and not to other cell-types such as amphibian heart, lung, cultured fibroblasts, etc. When fluorescent anti-keratin antibody was applied to the monolayer of cells which grew out from foot-web explants three days after explants were set up, all but 3 out of 6,700 cells (*i.e.* over 99.9%) bound the antibody. For the nuclear transfer experiments to be described, nuclei were taken from cells which had grown out from foot-web explants, three and one half days after they were set up. Therefore over 99.9% of these donor cells were keratinized at the time of nuclear transplantation.

Development of Nuclear-Transplant Embryos

Factors affecting the transplantation of nuclei from cultured cells in *Xenopus* have been described in detail by Gurdon and Laskey (1970a,b). The transplantation of nuclei from cultured cells is technically difficult and is attended by reasonable success only if serial nuclear transfers are carried out. Serial nuclear transplantation is believed to provide a better indication of the true developmental capacity of transplanted nuclei than first transfers, for the following reasons. A cultured cell divides very slowly and infrequently compared to cleavage cells, and in some cases, such as cultured skin cells, they do not divide at all. When the nucleus of a slow-dividing, or non-dividing, cell is transplanted to an egg, the egg will always divide according to its usual time schedule, that is at about two hours after fertilization or nuclear transplantation. Normally the egg and sperm nuclei should have completed replication within one and a half hours of fertilization. Nuclei from cultured cells start to replicate soon after transplantation but usually fail to complete chromosome replication before the first cleavage of egg cytoplasm. As a result, their incompletely replicated chromosomes enter only one of the first two cleavage blastomeres or may even be left in neither. A blastula is therefore formed which is a mosaic of cells with nuclei which are deficient in different ways for various chromosomes. It sometimes happens, however, that the transplanted nucleus passes undivided to only one of the first two blastomeres, and a partial blastula is formed with nucleated cells in one half and uncleaved cytoplasm in the other half. Under these conditions the transplanted nucleus has two S-phase periods instead of the usual one in which to replicate its chromosomes, and usually succeeds, it seems, in largely, if not fully, completing replication within this time. Thus the cells of a partially cleaved embryo are likely to have more normal chromosomes (that is, ones which are more representative of those of the originally transplanted nucleus) than cells of completely cleaved embryos. On the other hand the developmental capacity of the cells in a partially cleaved embryo cannot be revealed because half the embryo is uncleaved. Serial transplantation permits the nuclei of partially cleaved first-transfer embryos to form a complete blastula which is capable of developing as well as the genetic constitution of its chromosomes will permit. Some of these events are illustrated diagrammatically in Fig. 1. These are the reasons why our cultured cell nuclear transfer experiments always involve serial transfers.

The development obtained by the serial transplantation of "determined" skin cell nuclei is summarized in Table I. It is seen that the nuclei of determined skin cells promote development with the same success as the nuclei of cells grown out from other tissues or organs. In the case of the non-skin cells shown in Table I, the cells which grow out from cultured explants are thought to be fibroblasts which they resemble in morphology. Evidently there is no detectable difference in developmental capacity between the nuclei of fibroblast-like cells and those of cells determined for keratinization.

To test the developmental capacity of nuclei taken from fully differentiated cells, explants of foot-web skin were cultured under the "without plasma" conditions referred to above. The results of first and serial transfers of such nuclei are shown in Table II. There are two major conclusions from these experiments. The first is that the nuclei of differentiated cells promote embryonic development just as successfully as the nuclei of determined cells. The second important conclusion is that tadpoles composed of a wide range of different cell-types quite unrelated to skin, can be formed under the influence of

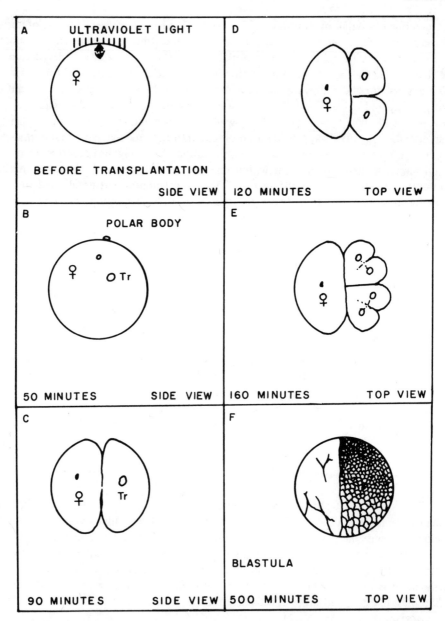

Figure 1. Sequence of stages believed to take place when partial blastulae are formed by the transplantation of cultured cell nuclei. ♀, egg pronucleus; Tr, transplanted nucleus. In this series of events, the transplanted nucleus fails to divide when the egg divides into the two-cell stage. One of the first two blastomeres receives only the ultraviolet irradiated egg-nucleus and does not divide further. The other blastomere divides at the normal times, with coordinated nuclear divisions. As a result, a partial half-cleaved blastula is formed, which cannot develop normally. Serial nuclear transfers are carried out by dissociating cells from the cleaved part of partial blastulae. In this way complete blastulae, capable of normal development, are prepared from a cultured cell nucleus.

nuclei which are mitotic products of a differentiated skin cell nucleus. This means that cell specialization does not involve any loss of, or permanent change in, genes. This applies to genes which are needed, for example, for the differentiation of nerve cells, muscle cells, lens cells, etc. Though genes of this kind will never be needed for skin cell differentiation, they are retained in a potentially active state in skin cells.

Validity of Nuclear Transfer Experiments

The principle of a nuclear transfer experiment is to eliminate the nucleus of an unfertilized egg, and to insert a transplanted nucleus in its place. In *Xenopus,* the egg nucleus is most conveniently eliminated by ultraviolet irradiation, and the procedure commonly used has proved very reliable. Nevertheless it is very desirable to have independent proof that the nuclei of each nuclear-transplant embryo are derived solely from the transplanted nucleus and not from a failure to eliminate the egg nucleus. For some

TABLE I

THE DEVELOPMENT OF NUCLEAR-TRANSPLANT EMBRYOS PREPARED FROM THE NUCLEI OF CELLS GROWN IN CULTURE FROM ADULT FROG ORGANS*

	Total number of nuclear transfers	% of total transfers forming partial or complete blastulae	% of serial transfer clones which contain	
			Muscular response tadpoles	Heart-beat tadpoles
Adult skin	1031	31	22	13
Adult kidney	789	26	12	9
Adult lung	502	24	25	11
Adult heart	273	22	40	0
Swimming tadpoles	3546	30	41	4

*Results are mainly taken from Laskey and Gurdon (1970).

years a nuclear marker has existed in *Xenopus* which can provide direct proof of this kind (Elsdale, Fischberg & Smith, 1958). Normally all ribosomal RNA genes are located at one place (the nucleolus organizer) on one chromosome of *Xenopus laevis*. The anucleolate mutation involves the loss of these genes (Wallace and Birnstiel, 1966) so that embryos homozygous for the mutation possess no ribosomal genes nor nucleoli. Such embryos are lethal. In heterozygous condition, nuclei have only one nucleolus (instead of the usual two), but have half the normal number of ribosomal genes, a dose which is sufficient to sustain normal development and growth. For nuclear transfer experiments, nuclei heterozygous for the anucleolate mutation are transplanted to enucleated eggs laid by wild-type females. If the resulting nuclear transplant embryos contain only nuclei which are mitotic products of the transplanted nucleus, all their nuclei will have only one nucleolus. Any nuclei with two nucleoli must have been derived from the egg nucleus. These expectations are specified on the assumption that all tadpole nuclei are diploid. If however the egg nucleus, which is haploid, were to participate in development without becoming doubled by fusion with a polar body or by fusion of two of its mitotic products, 1-nucleolated nuclei could be derived from the egg nucleus. Similarly 2-nucleolated nuclei could be derived from the transplanted nucleus

TABLE II

THE DEVELOPMENT OF NUCLEAR-TRANSPLANT EMBRYOS PREPARED FROM THE NUCLEI OF DIFFERENTIATED SKIN CELLS OF ADULT FROG FOOT WEBS
(From Gurdon *et al.*, 1975)

Total number of nuclear transfers	Partial blastulae	Partial blastulae used to make serial clones*	Serial clones which included:	
			Muscular response tadpoles	Heart-beat tadpoles
129	28	11	8	6
100%	22%	8%	6%	4½%

*Although 28 partial blastulae were obtained, it was possible within the time available to prepare serial clones from only 11 of these. There was no reason to doubt that the remaining 17 would have given serial transfer results comparable to that given by the 11 partial blastulae actually used. In this case the proportion of skin cell nuclei capable of promoting heart-beat development would have risen to over 10%.

if they are tetraploid.   However as long as the ploidy of nuclei in a nuclear transplant-embryo, as well as its nucleolar number, is known, there can be no doubt about the origin of these nuclei.

For these reasons, pieces of tissue from nuclear-transplant embryos are normally removed to prepare chromosome counts and nucleolar counts.   The results of some analyses of this kind are shown in Table III.   Results of this kind provide proof that the nuclei of a nuclear-transplant embryo are products of transplanted nuclei, and therefore these embryos provide valid tests of their developmental capacity.   Elsewhere we have reported tests which formally eliminate the remote possibility that ultraviolet light might change the genetically 2-*nu* to a genetically 1-*nu* condition (Gurdon & Laskey, 1970a).

TABLE III

CHROMOSOME AND NUCLEOLAR ANALYSES OF EMBRYOS OBTAINED BY THE SERIAL TRANSPLANTATION OF ADULT SKIN CELL NUCLEI

This table includes results from Gurdon and Laskey (1970a) and Gurdon *et al.* (1975). Of 89 tadpoles, all were clearly of transplant-nucleus, and not egg-nucleus, origin. There is no evidence for the participation of the egg-nucleus in nuclear transplant-embryo development.

Nuclear and nucleolar condition	Donor nucleus origin implied	Number of muscular response or heart-beat tadpoles
Haploid, 1-nucleolated	Egg nucleus	-
Diploid, 1-nucleolated	Transplant-nucleus	67
Diploid, 2-nucleolated	Egg nucleus (doubled)	-
Triploid, 2-nucleolated	Egg nucleus fused with transplant nucleus	-
Tetraploid, 2-nucleolated	Transplant-nucleus	22
Tetraploid, 4-nucleolated	Egg nucleus (quadrupled)	-

Changes in Gene Activity Which Follow Nuclear Transplantation

The fact that swimming tadpoles can be prepared by the transplantation of nuclei from specialized skin cells shows not only that unexpressed genes are retained by specialized cells, but also that the activity of these unexpressed genes can be restored. Evidently the muscle, nerve, and eye cells of an embryo prepared from a transplanted skin cell nucleus are not synthesizing keratin at the rate of keratinized skin cells, if indeed at all. Similarly myosin and crystallin genes which are presumed to have become inactive in skin cells can be brought back into activity by exposure to the cytoplasm of an egg or embryo. This is one of very few situations, if indeed there are any other circumstances, where genes which have been developmentally switched off, can be reproducibly reactivated. We would very much like to take advantage of this experimental situation to analyze how gene activity is changed. The mechanism which leads to the reactivation of unexpressed genes in keratinized skin cell nuclei after transplantation to eggs is almost certainly the same as that which brings about gene activation in the normal development of fertilized eggs. A sensitive way of monitoring the activity of keratin genes after nuclear transplantation will be to purify frog keratin mRNA, prepare labelled cDNA complementary to it, and ask whether the RNA made by the nuclear-transplant blastulae hybridizes with the cDNA to a greater extent than RNA made by blastulae grown from fertilized eggs. Until this is done, however, we have only some very general indications of the way by which genes may be controlled in early development.

One of the most striking events which accompanies early amphibian development is the accumulation, in cleavage nuclei, of proteins which have been synthesized in the cytoplasm. This effect is seen in nuclear transplant embryos as well as in fertilized eggs. The phenomenon was first observed by Arms (1968) and Merriam (1969) in *Xenopus*. Ecker and Smith (1971) have shown in *Rana pipiens* that many of the proteins synthesized during oocyte maturation become concentrated in the nuclei of blastula cells. As yet we have no information about the identity or function of these proteins. It is, however, a reasonable guess that some of them are associated with chromosomes, and that their function includes that of regulating gene activity. A classical concept of development attributes great importance to the unequal distribution of materials in an egg. If different egg substances become partitioned into different cleavage cells, this could account for the first differences which arise between regions of an embryo, and in this way for the first steps of cell differentiation. Clearly, cytoplasmic proteins which are synthesized during oogenesis or soon after fertilization and which subsequently accumulate in the nuclei of blastulae are strong candidates for these egg substances whose unequal distribution is of morphogenetic importance.

One approach to the investigation of these cytoplasmic proteins which accumulate in nuclei is to ask how specific their nuclear affinity is. Is there some special mechanism by which certain proteins become nucleus-associated, or are all proteins accumulated in cleavage nuclei to the same extent? Some years ago it was found that labelled histones, injected into oocytes or eggs, are concentrated in nuclei to a spectacular extent, whereas other proteins not normally present in nuclei such as bovine serum albumin or ferritin are not (Gurdon, 1970) (Table IV). Recently Bonner (1975) in this laboratory has extended these observations in two interesting ways. First he has tested a wide range of proteins for their ability to accumulate in oocyte nuclei. He found that, of those tested, only histones show a nuclear concentration, and that other molecules of similar size and overall charge, such as lysozyme, do not. The second type of experiment carried out by Bonner was to label oocytes by incubation in [^{35}S] methionine, so that both nucleus and cytoplasm contain labelled proteins. He then injected labelled nuclear proteins into the cytoplasm of one set of unlabelled oocytes and labelled cytoplasmic proteins into the cytoplasm of another set of unlabelled oocytes. The fate of the injected labelled proteins was then followed by autoradiography of sectioned oocytes and by scintillation counting of manually separated nuclei and cytoplasms. The results clearly showed that within a few hours the nucleus-derived proteins had migrated back into the oocyte nuclei, whereas the cytoplasmic proteins had remained localized in the cytoplasm of the injected cells.

Prospects

The experiments outlined above, on the injection of nuclear proteins into oocytes and eggs, indicate that the accumulation of proteins in nuclei is a specific process. They also demonstrate that proteins can be removed from a cell and reinserted into another cell without losing the properties responsible for their special intracellular distribution.

TABLE IV

SELECTIVE ACCUMULATION OF [^{125}I]-LABELLED PROTEINS IN THE NUCLEI OF INJECTED XENOPUS EGGS

	Average number of autoradiographic grains per area of section		
	Transplanted nucleus	Egg pronucleus*	Egg cytoplasm
[^{125}I] histone	∿ 200	∿ 200	81 $\pm$ 14
[^{125}I] bovine serum albumin	44 $\pm$ 10	38 $\pm$ 7	102 $\pm$ 13

*The egg nucleus has been ultraviolet irradiated, but has not disintegrated at the time the nuclear-transplant eggs were fixed, that is 50 min after nuclear transplantation.

This means that we may hope to be able, eventually, to isolate developmentally important nuclear proteins, fractionate them, and reintroduce them into cells in such a way that they retain their biological activity. Over the last several years evidence has been accumulating that, to a very encouraging extent, macromolecules can be extracted from cells, quite extensively purified, and yet function normally when reintroduced into other living cells, a subject which has been reviewed elsewhere (Gurdon, 1974b). The fact that experiments of this type can be done opens up a way of identifying the function of purified macromolecules in general, and in particular of nuclear proteins.

REFERENCES

Arms, K. (1968). *J. Embryol. exp. Morph.* 20, 367.

Bonner, W.M. 1975). *J. Cell Biol.* 64, 421.

Ecker, R.E. and Smith, L.D. (1971). *Devel. Biol.* 24, 559.

Elsdale, T.R., Fischberg, M. and Smith, S. (1958). *Exptl. Cell Res.* 14, 642.

Elsdale, T.R., Gurdon, J.B. and Fischberg, M. (1960). *J. Embryol. exp. Morph.* 8, 437.

Gurdon, J.B. (1962). *J. Embryol. exp. Morph.* 10, 622.

Gurdon, J.B. (1970). *Proc. R. Soc. B.* 176, 303.

Gurdon, J.B. (1974a). *The Control of Gene Expression in Animal Development* (Harvard Press), p. 160.

Gurdon, J.B. (1974b). *Nature* 248, 772.

Gurdon, J.B. and Laskey, R.A. (1970a). *J. Embryol. exp. Morph.* 24, 227.

Gurdon, J.B. and Laskey, R.A. (1970b). *J. Embryol. exp. Morph.* 24, 249.

Gurdon, J.B., Woodland, H.R. and Lingrel, J.B. (1974). *Devel. Biol.* 39, 125.

Gurdon, J.B., Laskey, R.A. and Reeves, O.R. (1975). *J. Embryol. exp. Morph.* 34. (In press.)

Laskey, R.A. and Gurdon, J.B. (1970). *Nature* 228, 1332.

Merriam, R.W. (1969). *J. Cell Sci.* 5, 333.

Reeves, O.R. (1975). *J. Embryol. exp. Morph.* 34. (In press.)

Reeves, O.R. and Laskey, R.A. (1975). *J. Embryol. exp. Morph.* 34. (In press.)

Wallace, H.R. and Birnstiel, M.L. (1966). *Biochim. Biophys. Acta.* 114, 296.

Woodland, H.R., Gurdon, J.B. and Lingrel, J.B. (1974). *Devel. Biol.* 39, 134.

PLANT TISSUE CULTURE METHODS IN SOMATIC HYBRIDIZATION BY PROTOPLAST FUSION AND TRANSFORMATION *

Oluf L. Gamborg

Prairie Regional Laboratory
National Research Council
Saskatoon, Saskatchewan S7N0W9 Canada

Introduction

Current concepts of gene structure and regulation were formulated largely on the basis of biochemical and molecular genetical analyses of prokaryote cells. Their rapid growth makes it possible to obtain large quantities of single cells and measure biochemical and physiological events relatively quickly. It is not surprising that microorganisms became the material of choice for investigations on the metabolic and molecular genetic regulation of growth and differentiation in living cells.

Parallel with the expanding knowledge of molecular genetics and cell regulation came discoveries of transfer of genetic material in microorganisms. This process can occur through uptake of DNA (transformation), by transduction involving viral carriers, or by conjugation (Merrill and Stanbro, 1974).

By using phages and plasmids, it is feasible to construct and transfer specific genes from one organism to another, where the genetic material may integrate and become functional (Cohen *et al.*, 1973).

Similar genetic manipulations are now being envisaged with plant cells. The procedures focus on DNA-mediated transformation and hybridization by cell fusion and involve methods of plant cell culture (Nickell and Heinz, 1973; Gamborg and Miller, 1973; Gamborg *et al.*, 1975).

Plant Cell Culture and Morphogenesis

Plant cells can be grown indefinitely as masses on agar (callus) or in liquid suspension

---

*NRCC No. 14691.

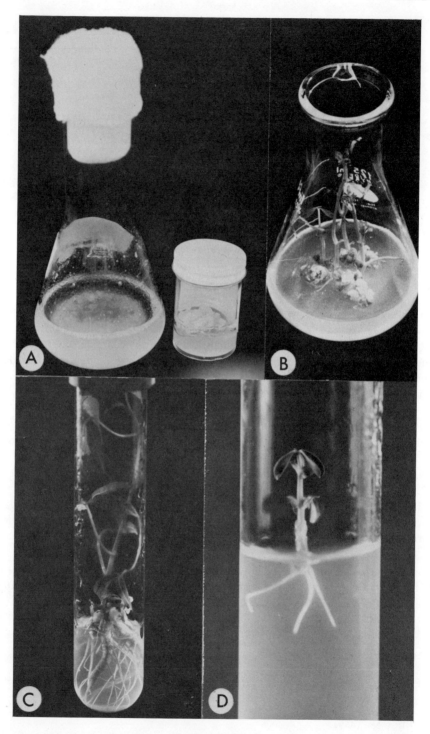

(Fig. 1A), and produced in any desired quantity. Generally the cells require only mineral salts, sucrose, and certain vitamins and growth hormones (Street, 1973). The latter are needed to induce cell division.

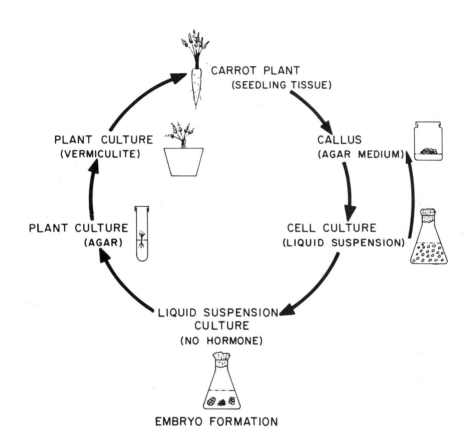

Figure 2. Plant cell culture. Callus, cell suspension culture and plant regeneration. Outline of tissue culture and embroygenesis in carrot cells.

⟨————————

Figure 1. Plant tissue culture and morphogenesis.
  A. Cell suspension and callus culture of pea *(Pisum sativum)*.
  B. Plant regeneration from pea shoot tip callus.
  C. Cassava plantlet produced by meristem culture.
  D. Pea plantlet grown from shoot tip.

Cells grown in suspension culture appear to be undifferentiated and generally do not show any structural or biochemical evidence for functions other than continued division. However, totipotency may be expressed. When produced and cultured under suitable conditions, the cells can be induced to differentiate, and regenerate complete plants (Street, 1973; Murashige, 1974) (Fig. 2).

The process of morphogenesis may occur by embryogenesis in the absence of exogenous hormones or by organogenesis induced by cytokinins. The cytokinins are adenine-derivatives, which stimulate shoot formation in most plants except cereals and grasses. Shoots may be readily obtained from stem sections, and can be produced from callus grown in the presence of cytokinins (Kartha *et al.,* 1974a; Gamborg *et al.,* 1974a) (Fig. 1B, C). When shoots are excised and transferred to other media, rooting can be achieved (Fig. 1D).

There are numerous examples of plant regeneration from callus, and in some cases the techniques can be utilized as a routine method for vegetative propagation (Earle and Langhans, 1974; Kartha *et al.,* 1974b; Murashige, 1974) (Fig. 3).

Apart from hormones and tissue origin, the culture medium and light and temperature play a significant role in the success of plant regeneration (Murashige, 1974). The specific conditions must be determined for each plant species.

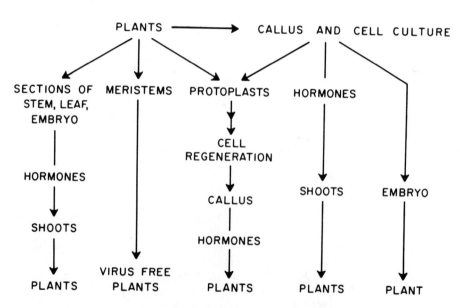

Figure 3. Outline of protoplast cell culture, and methods of morphogenesis.

Plant Protoplast Isolation and Culture

Significant advances have been made in the technology for isolating plant proto-plasts (Cocking, 1973; Tempe, 1972; Gamborg *et al.,* 1973). They are released by incu-bating the tissues with hydrolytic enzymes (cellulases, hemicellulases, pectinases) in solu-tions containing osmotic stabilizers such as hexitols or glucose (Fig. 4A). Protoplasts can be prepared in high yields within 6-12 hours depending upon the tissue (Kao *et al.,* 1971; Tempe, 1973; Constabel, 1975). The most common materials are leaf tissues and cells from suspension cultures, but they can be obtained from fruits, root and shoot materials (Nickell and Heinz, 1973).

Isolated protoplasts can be cultured and in some cases be induced to divide (Table I; Fig. 4B). Growing protoplasts in droplets has been very successful (Kao *et al.,* 1971). The nutritional requirement of protoplasts is similar to those of plant cells, but initially they often have special needs for sugars, calcium, glutamine or hormones. The first divisions may occur within 24-72 hours and a callus would subsequently be produced.

Cultured protoplasts reform a cell wall which in composition and structure resem-bles the original wall. (Albersheim, private communication; Fowke *et al,* 1974).

TABLE I

EXAMPLES OF PROTOPLASTS IN WHICH CELL
REGENERATION AND DIVISION HAS BEEN OBSERVED

Systematic	Common Name	Cell Origin
Ammi visnaga		culture
Bromus inermis	Brome grass	culture
Cicer arietinum	Chick pea	leaf
Brassica napus	Rapeseed	culture, leaf
Daucus carota	Carrot	culture, leaf
Glycine max	Soybean	culture
Linum usitatissimum	Flax	leaf
Medicago sativa	Alfalfa	leaf, culture
Melilotus alba	Sweet clover	leaf
Phaseolus vulgaris	Bean	leaf, culture
Pisum sativum	Pea	leaf
Pisum sativum	Pea	culture, shoot tip
Vicia hajastana		culture
Vigna sinensis	Cow pea	leaf

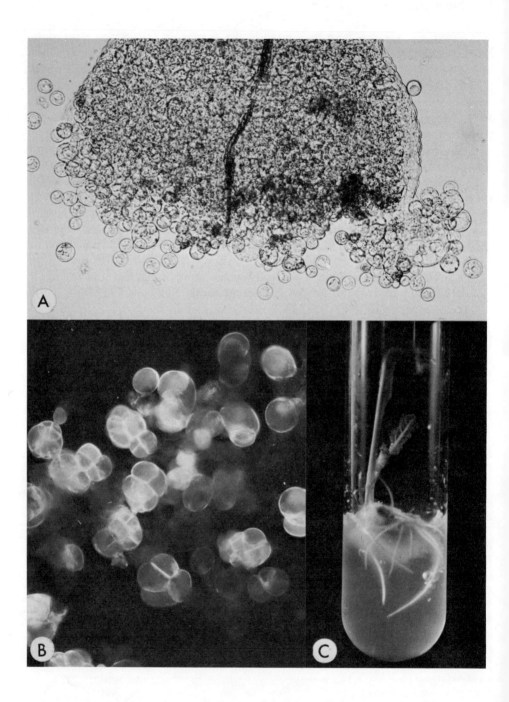

Mitosis and division on the reconstituted cells resembles in every respect that of cells grown in suspension culture (Fowke *et al.*, 1974). Although sustained division has been possible in protoplasts of some species, the lack of success in many others remains a serious handicap and warrants continued study (Eriksson *et al.*, 1975).

Genetic Modifications in Plants

Various methods are being explored to develop procedures which permit gene transfer and modification in somatic cells of plants (Fig. 5). Apart from their potential value in genetic analyses, these methods may provide a means for achieving wide crosses in plants. The natural barriers which prevent sexual crosses between plant genera normally do not exist when somatic cells are hybridized. Two methods have been considered for gene transfer: DNA-mediated transformation and cell fusion. Protoplasts are required for cell fusion, while transformation can be investigated also with other plant materials. Transformation or fusion with protoplasts involves four steps (Fig. 5):

1. **Protoplast isolation and culture**
   Develop procedures for isolating viable protoplasts, and establish methods for culturing the protoplast and regenerating dividing cells.
2. **Fusion/DNA-uptake**
   Carry out fusion/DNA-uptake.
3. **Cell selection and culture**
   Reconstitute the cells and select the hybrid/transformed cells.
4. **Plant production**
   Culture the genetically modified cells, under conditions appropriate for differentiation and plant regeneration.

Protoplast Fusion

Fusion can occur spontaneously between adjacent protoplasts during isolation. The process appears to commence with dilation of plasmadesmata connecting plant cells. Fusion of protoplasts from different sources must be induced. The isolated protoplasts are spherical, and an agglutinating agent is required to increase the area of membrane contact to facilitate fusion.

⟨————————

Figure 4. Protoplasts
   A. Isolation from shoot tips of pea seedling.
   B. Division in regenerated cells from rapeseed leaf protoplasts.
   C. Plantlet grown from callus derived from rapeseed leaf protoplasts.

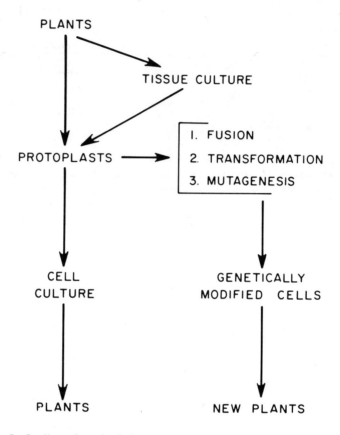

Figure 5. Outline of methods for genetic modification in somatic plant cells.

Although several compounds such as nitrate, lysozyme and treatments involving sudden osmotic changes and centrifugal forces were reported to facilitate fusion, the rates were very low (Eriksson *et al.*, 1975). Hartmann *et al.* (1973) prepared antisera and achieved effective agglutination of protoplasts from different plant genera, but fusion did not occur. Treatment at alkaline pH in the presence of calcium induced fusion of protoplasts from two tobacco mutants (Keller and Melcher, 1973) and interspecies hybrid tobacco plants were produced (Melcher *et al.*, 1974).

An effective method for protoplast fusion employing polyethylene glycol (PEG) has been introduced recently (Kao and Michayluk, 1974; Eriksson *et al.*, 1975). The polymer facilitates fusion of protoplasts of different plant genera. Protoplasts aggregate immediately when exposed to high concentrations of PEG (Fig. 6A, B). Fusion occurs when the PEG is eluted out and heterokaryons are formed (Fig. 4C) (Kao *et al.*, 1975; Fowke *et al.*, 1975). The fusion rate depends upon protoplast quality, molecular size of the PEG, and is also influenced by compounds and treatments which modify the plasma membrane (Kao *et al.*, 1975; Constabel and Kao, 1974).

Intergeneric fusion can be achieved consistantly at rates between 10 and 40% of the surviving protoplasts. There are no apparent difficulties in fusing protoplasts of widely different plant genera and families, and the tissue origin of the protoplasts does not seem to be critical for heterokaryon formation and division (Table II). Present evidence indicates that if at least one of the parental protoplasts normally can be induced to divide, there is a greater chance for division and hybrid cell formation after fusion. It is therefore essential that conditions for the culture and division of at least one of the parent protoplasts be developed before any attempts are made to obtain cell progeny from fused protoplasts. Leaf mesophyll protoplasts exhibit relatively low frequencies of division. However, after fusion with protoplasts from the PRL soybean cell culture line, the resulting heterokaryocytes divide. The stimulus for division seems to reside in the soybean protoplasts (Fig. 6D).

When heterokaryocytes produced by fusion divide, most of the leaf protoplasts deteriorate and, initially, the hybrids are the only cells with green chloroplasts (Table II). The hybrid cells contain a single nucleus, and it is apparent that nuclear fusion occurs prior to or during mitosis. The dividing cells contain chromosomes from both parental species (Fig. 6E).

Recognition of the hybrid cells based on the presence of green chloroplasts has been adequate as an aid in demonstrating the formation of intergeneric hybrids. Procedures for eliminating both parental cell species and the isolation of hybrids is required to obtain pure cultures of the hybrid lines. Selective culture media or mechanical isolation by micromanipulation may be feasible methods. Other systems include the use of metabolic mutants, toxins, antimetabolites or other selective chemicals (Chaleff and Carlson, 1974).

Transformation

Transformation may provide an alternate approach for achieving exchange of genetic information in plants. The process involves donor-DNA uptake, integration, replication and expression of new genetic information in the recipient organism or cell (Fig. 7) (Ledoux, 1971). The desirable result is the stable phenotypic expression of the genetic information coded for by the donor DNA. While cell fusion necessitates the use of protoplasts, investigations on transformation also have been pursued with plant organs. Using seeds or plant organs has the advantage that complete plant formation is ensured after introducing the DNA. Ledoux and colleagues have investigated extensively the uptake and fate of bacterial DNA in plants (Ledoux, 1971). They reported successful correction of biochemical mutants of *Arabidopsis thaliana* (L. Heyhn) requiring thiamine, thiazole or pyrimidine, after treatment of the seeds with DNA from microorganisms. The radioactive DNA fed to imbibing seeds was taken up and a portion stored in the cotyledons. The DNA was later translocated to the flower, and biochemical evidence was presented for a hybrid DNA in the F_1 progeny. Some of the transformed plants responded to the addition of thiamine to the growing medium, indicating incomplete correction (Ledoux, 1974).

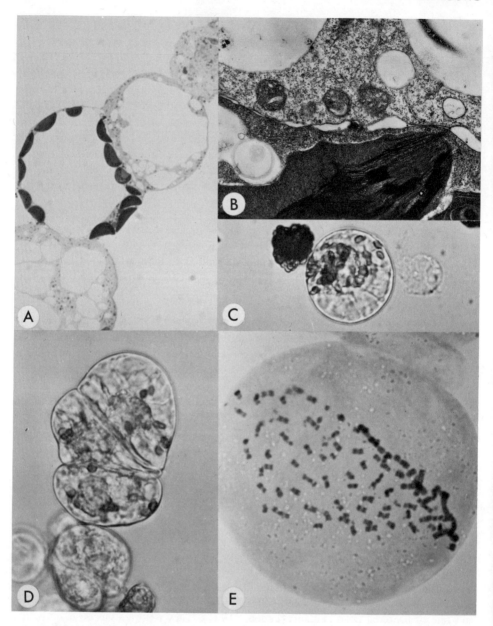

Figure 6.   Protoplast fusion.
A. Aggregate of protoplasts from pea leaf and *Vicia hajastana* cell cul-
ture produced in polyethylene glycol.
B. Ultrastructural section of A.
C. Fusion product (heterokaryocyte) of barley + soybean protoplasts.
D. Cells from barley-soybean heterokaryocyte.
E. Heterokaryon of corn-soybean in mitosis.

TABLE II

**INTERGENERIC PLANT PROTOPLAST FUSION
AND HETEROKARYOCYTE DIVISION***

SOURCE OF PROTOPLASTS

Leaf mesophyll		Cell Culture
Barley *(Hordeum vulgare)*	x	soybean *(Glycine max)*
Pea *(Pisum sativum)*	x	*Vicia hajastana*
Corn *(Zea mays)*	x	soybean *(Glycine max)*
Pea *(Pisum sativum)*	x	soybean *(Glycine max)*
Rapeseed *(Brassica napus)*	x	soybean *(Glycine max)*
Alfalfa *(Medicago sativa)*	x	soybean *(Glycine max)*
Sweet clover *(Melilotus alba)*	x	soybean *(Glycine max)*
Chick pea *(Cicer arietinum)*	x	soybean *(Glycine max)*
Angelica archangelica	x	carrot *(Daucus carota)*

*References: Constabel and Kao, 1974; Kao and Michayluk, 1974; Kao *et al.,* 1975; Kartha *et al.,* 1974c; Fowke *et al.,* 1975; Dudits, unpublished.

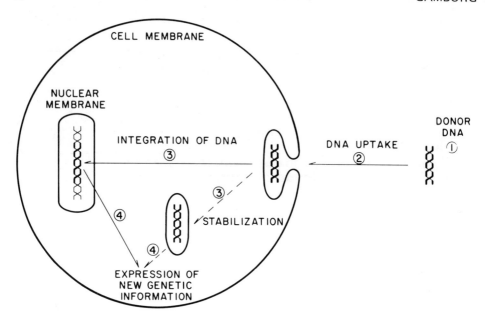

Figure 7. Model of DNA-mediated transformation in plants.

Hess (1972) observed a genetically stable change in flower color in progeny of *Petunia* after feeding DNA from a red-flowering (wild type) to plants of a white-flowering anthocyanin-mutant. Several molecular genetic models have been proposed to account for these observations (Holl *et al.*, 1975; Ledoux, 1971).

Cultured cells and particularly protoplasts may provide a more effective material. Protoplasts from cell suspension cultures of soybean, carrot and *Ammi* take up bacterial DNA (Ohyama *et al.*, 1972). Up to 1% of the radioactive DNA was incorporated into TCA-insoluble fractions during a 4-hour incubation period. The bacterial DNA after uptake appears inside the protoplasts as the double-stranded form (Ohyama, 1975).

The protoplasts reform a cell wall and divide after the DNA treatment. Plant DNA supplied to Petunia protoplasts has been observed to enter the nucleus (Hoffman and Hess, 1973). Although protoplasts may be highly desirable as materials for transformation experiments, their usefulness is restricted at present because cell regeneration and division at high frequency is limited to a few species. Moreover, biological marker systems for selection of transformed cells are not generally available.

The capability of plant cells to utilize new carbon sources should have potential as a selection system. Plant cells rarely grow on mannitol or lactose although these compounds are separated from the glycolytic pathway by only one or two enzymes. DNA from bacteria which utilize these compounds has been fed to soybean protoplasts, and cells were obtained which grew on lactose-mannitol media. The growth rates were extremely low and biochemical analyses have not been possible.

An alternative approach to feeding isolated DNA involves the use of transducing phages as carriers (Doy et al., 1973). Specific genes may be carried by phages which then serve as vectors to facilitate the introduction of DNA into the plant cells (Merrill and Stanbro, 1974).

Although controversies exist about classifying the present observations as genetic transformation, there is substantial evidence for DNA uptake and integration into the plant genome. However, considerable uncertainty still exists about the mode of integration and the stability of the donor DNA in the host cells.

Mutant Cells

The plant cell and protoplast systems can be adapted and utilized for the purpose of mutant cell production. Cells from suspension cultures can be exposed to chemical or physical mutagens followed by selection for desirable mutants (Chaleff and Carlson, 1974).

The isolation of metabolic mutants of plant cells from the 'wild' type is not as readily achieved as with bacteria. The procedure of 5-bromodeoxyuridine BUdR) treatment followed by light exposure can be employed. Actively growing cells on minimal media incorporate BUdR and are destroyed while slow-growing mutants survive (Kao and Puck, 1968; Chaleff and Carlson, 1974).

Mutants with nucleic acid base analogs or antibiotic resistance may be effective, because a positive selection system for eliminating the normal, sensitive cells can be employed. The same principle applies to the isolation of cells resistant to amino acid analogs and certain pathogenic toxins.

Ohyama has employed protoplasts from soybean cell cultures to obtain BUdR-resistant mutants (Fig. 8). The mutants grow rapidly in liquid suspension in the presence of BUdR, which is incorporated into the DNA (Ohyama, 1974).

At present the most readily available plant material consists of diploid cells. For dominant or organelle residing characteristics such cells may be adequate, but haploid cells would have wide application in mutagenic studies. Haploid cells are becoming readily available through anther and pollen culture (Smith, 1974). In some plant species, anthers (pollen) can be cultured on defined media, where they form embryos and complete (haploid) plants (Sunderland, 1971). Cells obtained from anthers may not remain haploid but undergo chromosome changes to diploid and aneuploids.

Since haploid plants can be obtained, leaf mesophyll protoplasts from such plants could be used as materials for mutagenesis.

Hybrid Plants

Cells obtained after fusion or transformation can be subjected to conditions for inducing plant regeneration. It is imperative, however, to determine the specific requirements for inducing organogenesis in cells regenerated from the particular protoplasts employed in the genetic experiment. Plants have been obtained from protoplasts of several species including carrot (Grambow *et al.*, 1972), rapeseed (Fig. 4C) (Kartha *et al.*, 1974a, b, c) and from cells produced by fusion of protoplasts from two tobacco species (Carlson *et al.*, 1972; Melcher *et al.*, 1974). The latter constituted interspecies hybrid plants. Similar plant regeneration may be anticipated from intergeneric somatic hybrid cells. In view of the ease with which hybrid cells from widely different plants can be produced and induced to divide, the process of hybrid plant formation may be realized when suitable culture conditions have been developed.

Figure 8.   Selection for BUdR-resistant soybean cells regenerated from protoplasts treated with mutagens (Ohyama, 1975).
A. Control. Growth in absence of BUdR.
B. Colonies of BUdR-resistant cells.

Concluding Discussion and Applications

Many advances in the technology of tissue culture and the handling of protoplasts have now made it possible to investigate genetic manipulation of somatic plant cells. The development of a new and efficient technique has resulted in fusion of protoplasts and the production of intergeneric hybrids between cells of various cultivated plants. The results show that the natural barriers which prevent sexual crosses between plant genera are circumvented in the fusion and growth of somatic hybrid cells.

Hybrid cells can be recognized by the presence of green chloroplasts from the parental leaf protoplasts. Such a detection system is made possible by designing conditions in which the unfused parental leaf protoplasts fail to survive. Appropriate selection systems which permit isolation of hybrids and elimination of all parental cells are being developed. It will then be possible to obtain clones and determine the stability of the karyotypes, which consist of a combination of the parental chromosomes, and utilize the materials for biochemical and genetic analyses.

The evidence for transformation in plants is restricted largely to correction of biochemical lesions with bacterial DNA. The practical value of the technique, however, may be limited unless plant DNA can be used. Progress in the testing and development of gene transfer procedures depends upon the careful choice of suitable plant material such as the use of isogenic lines with single gene differences expressed as a clear cut phenotypic variation.

There is no information on the role of competence for DNA-mediated transformation in plants. The relative importance of molecular size, purity and degree of DNA denaturation has not been ascertained. An assessment of these problems becomes feasible with genetically characterised systems in which the encoded information is expressed in a relatively short time.

Present techniques also permit investigation of the uptake of nuclei, plastids and other organelles by protoplasts (see review by Eriksson *et al.*, 1975). Organelle transfer to protoplasts and subsequent reconstitution of the cells may provide an elegant approach for genetic analyses and for investigations on gene control of biochemical and developmental processes in higher plants. Such systems may be helpful in elucidating the control of morphogenesis, a phenomenon which is very readily achieved in cells of such plants as tobacco or carrot, but not in cells of most cultivated plants (*e.g.* soybean, corn).

As hybrids and genetically modified cells become available, it is expected that methods can be designed for inducing differentiation and plant regeneration. Although the capability for totipotency in somatic cells has been adequately documented, attempts to elicit the manifestation of the process are frequently unsuccessful. It is imperative to determine the necessary conditions and requirements for plant regeneration in one of the parental lines before attempting hybrid plant production.

Present progress in tissue culture and genetic manipulation procedures makes it possible to direct their application to problems of practical importance (Nickell and Heinz, 1973). Methods for producing wide crosses are required for plant breeding and crop improvement (Borlaug, 1971; Wittwer, 1974). Existing methods which depend upon sexual crosses are limited to closely related species. The somatic cell fusion and transformation method would make it possible to obtain an infinite variety of crosses. The products of these combinations may then be employed in conventional plant breeding systems. Somatic cell genetics, thus, provides the means to transfer desirable characteristics between plants of different families.

The objectives of more effective disease resistance, greater tolerance to extreme environments and plants with improved protein quality and quantity depend on such techniques (Wittwer, 1974).

The need for novel genetic procedures is suggested by more distant goals of significant agronomic potential. Current objectives include plans to introduce greater photosynthetic efficiency into crop plants. The product of photosynthesis of some plants are C-3 compounds, in others it is a C-4 compound. In the latter type the photosynthetic process is more efficient because the photorespiration is lower. The sites of genetic control and regulation of these properties have not been established.

Another objective of great importance is directed towards equipping non-leguminous plants with the ability to fix nitrogen from the air. Nitrogen fixation in legumes such as soybean and peanuts is a symbiotic process involving the plant and a *Rhizobium* species. Fixation takes place in nodules formed on the roots. The process is determined by bacterial and plant genes.

Although skepticism exists about the feasibility of manipulating the complex biochemical processes of photosynthesis and nitrogen fixation, somatic cell genetics may provide a tool to investigate these objectives. The design of less complicated.systems to ascertain the genetics of separate steps of these processes should be the primary consideration (Holl *et al.*, 1975).

The predictions of an imminent shortage of food may be alleviated by raising the efficiency of production and improving the yields of existing crop types. However, in view of continued population increases the urgency for new crops with superior productivity, growth efficiency and product quality remains a vital and compelling objective, which can be realized only through methods of somatic cell hybridization.

Acknowledgements

I am grateful to the following colleagues and associates in Saskatoon for many helpful discussions and for providing data and photographic materials: Drs. F. Constabel, L. Fowke (University of Saskatchewan), B. Holl, K. Kartha, K. Kao, K. Ohyama and L. Pelcher. I thank Mr. A. Lutzko for preparing the drawings and photographic plates.

REFERENCES

Borlaug, N.E. (1971). *Cereal Science Today 16*, 401.

Carlson, P.S., Smith, H.H. and Dearing, R.D. (1972). *Proc. Nat. Acad. Sci. USA 69*, 2292.

Chaleff, R.S. and Carlson, P.S. (1974). *Am. Rev. Gen. 8*, 267.

Cocking, E.C. (1972). *Annu. Rev. Plant Physiol. 23*, 29.

Cohen, S.N., Chang, A.C.Y., Boyer, H.W. and Helling, R.B. (1973). *Proc. Nat. Acad. Sci. USA 70*, 3240.

Constabel, F. (1975) in *Plant Tissue Culture Methods*, eds. Gamborg, O.L. and Wetter, L.R. (Saskatoon: National Research Council of Canada).

Constabel, F. and Kao, K.N. (1974). *Can. J. Bot. 52*, 1603.

Doy, C.H., Gresshoff, P.M. and Rolfe, B. (1973) in *The Biochemistry of Gene Expression in Higher Organisms*, eds. Pollak, J. and Lee, J. Wilson (Australia and New Zealand Book Co.), pp. 21-37.

Earle, E.D. and Langhans, R.W. (1974). *J. Amer. Soc. Hort. Sci. 99*, 128.

Eriksson, T., Bonnett, H., Glimelius, K. and Wallin, A. (1975) in *Advances in Plant Science Through Tissue Culture*, eds. Street, E.H. and Cocking, E.C.

Fowke, L.C., Bech-Hansen, C.W., Constabel, F. and Gamborg, O.L. (1974). *Protoplasma 81*, 189.

Fowke, L.C., Rennie, P.J., Kirkpatrick, J.W. and Constabel, F. (1975). *Can. J. Bot. 53*, 272.

Gamborg, O.L. and Miller, R.A. (1973). *Can. J. Bot. 51*, 1795.

Gamborg, O.L., Kao, K.N., Miller, R.A., Fowke, L.C. and Constabel, F. (1973). *Coll. Int. Cent. Nat. Rech. Sci. 212*, 155.

Gamborg, O.L., Constabel, F. and Shyluk, J.P. (1974). *Physiol. Plant. 30*, 125.

Gamborg, O.L., Constabel, F., Fowke, L., Kao, K.N., Ohyama, K., Kartha, K. and Pelchen, L. (1974). *Can. J. Genet. Cytol. 16*, 737.

Grambow, H.J., Kao, K.N., Miller, R.A. and Gamborg, O.L. (1972). *Planta 103*, 348.

Hartmann, J.X., Kao, K.N., Gamborg, O.L. and Miller, R.A. (1973). *Planta 112*, 45.

Hess, D. (1972). *Naturwissenschaften 59*, 348.

Hoffman, F. and Hess, D. (1973). *Z. Pflanzenphysiol. 69,* 81.

Holl, F.B., Gamborg, O.L., Ohyama, K. and Pelcher, L. (1974) in *Advances in Plant Science Through Tissue Culture,* eds. Street, E.H. and Cocking, E.C.

Johnson, C.B. and Grierson, D. (1974) in *Current Advances in Plant Science,* ed. Smith, H., Vol. 2, No. 9, Commentaries (Maxwell, Oxford), pp. 1-12.

Kao, F.T. and Puck, T.T. (1968). *Proc. Nat. Acad. Sci. USA 60,* 1275-1281.

Kao, K.N., Gamborg, O.L., Miller, R.A. and Keller, W.A. (1971). *Nature New Biology 232,* 124.

Kao, K.N. and Michayluk, M.R. (1974). *Planta 115,* 355.

Kao, K.N., Constabel, F., Michayluk, M.R. and Gamborg, O.L. (1974). *Planta 120,* 215.

Kartha, K.K., Gamborg, O.L., Constabel, F. and Kao, K.N. (1974a). *Can. J. Bot. 52,* 2435.

Kartha, K.K. Gamborg, O.L., Constabel, F. and Shyluk, J.P. (1974b). *Plant Sci. Lett. 2,* 107.

Kartha, K.K., Michayluk, M.R., Kao, K.N., Gamborg, O.L. and Constabel, F. (1974c). *Plant Sci. Lett. 3,* 265.

Keller, W.A. and Melchers, G. (1973). *Z. Naturforschung 28,* 737.

Ledoux, L. [ed.] (1971). *Informative Molecules in Biological Systems* (Amsterdam: North-Holland Publ. Co.).

Ledoux, L., Huart, R. and Jacobs, M. (1971). *Eur. J. Biochem. 23,* 96.

Ledoux, L., Huart, R., and Jacobs, M. (1974). *Nature 249,* 17.

Melchers, G., Keller, W. and Labib, G. (1974). *Proc. Cong. Plant Tissue and Cell Culture* [Abstracts] (Leicester), p. 147.

Merrill, C.R. and Stanbro, H. (1974). *Z. Pflanzenphysiol. 72,* 371.

Murashige, T. (1974). *Annu. Rev. Plant Physiol. 25,* 135.

Nickell, L.G. and Heinz, D.J. (1973) in *Genes Enzymes and Populations,* ed. Srb, A.M. (New York: Plenum), pp. 109-128.

Ohyama, K., Gamborg, O.L. and Miller, R.A. (1972). *Can. J. Bot. 50,* 2077.

Ohyama, K. (1974). *Expt. Cell. Res. 89,* 31.

Ohyama, K. (1975) in *Plant Tissue Culture Methods,* eds. Gamborg, O.L. and Wetter, L.R. (Saskatoon: National Research Council of Canada).

Smith, H.H. (1974). *BioScience 25(5),* 269.

Street, H.E. (1973) in *Botanical Monographs,* Vol. II (London: Blackwell Scientific Publications).

Sunderland, N. (1971). *Sci. Prog. 59,* 527.

Tempe, J. [ed.] (1973) in *Protoplastes et Fusion de Cellules Somatique Vegetales, Coll. Int. Cent. Nat. Rech. Sci.* (Paris), p. 212.

Wittwer, S.H. (1974). *BioScience 24(4),* 216-224.

HORMONE MEDIATED INTEGRATION OF SEEDLING PHYSIOLOGY

J.E. Varner

Biology Department
Washington University
St. Louis, Missouri 63130

Introduction

The mobilization of the endosperm reserves of barley seeds (and of many other cereal grains) during the early growth of the seedling is initiated and controlled by gibberellins from the embryo. The gibberellins are synthesized in the embryo tissues after germination begins and have as their principal target tissue the cells of the aleurone layers (Yomo and Varner, 1971). These aleurone cells produce and secrete the hydrolases that hydrolyze the starch and protein reserves of the endosperm. The end products of this hydrolysis are absorbed by the scutellum and, after limited metabolism, transported to the growing regions of the embryo.

In barley the aleurone tissue is three cell layers thick. During normal germination those aleurone cells nearest the scutellum are the first to receive the gibberellins and the first to respond. As early growth of the embryo proceeds, cells farther and farther from the scutellum receive and respond to the gibberellins and secrete the hydrolases that digest the starchy endosperm underlying them.

Mobilization of the endosperm is not required for germination and, although gibberellins can enhance germination of the seeds of many species, the effect of the gibberellins is directly on the embryo and does not involve mobilization of endosperm reserves (Chen and Park, 1973).

Because the aleurone layers are the principal target for the gibberellins and because these layers are easily peeled away from the starchy endosperm, isolated aleurone layers of barley and wheat have been widely used to study the response of aleurone cells to exogenous hormones. Production of the α-amylase, of the protease, of part of the ribonuclease and of part of the β-glucanase has been shown to be due to synthesis of these enzymes in response to the added hormone. In response to the added gibberellins, barley aleurone layers release into the medium (in addition to the amylase, protease, β-glucanase and ribonuclease) increased amounts of phosphatase, pentosanase, peroxidase, esterase and glucosidase (see Jacobsen and Knox, 1974). It is not known whether the appearance of these enzymes in the medium is due solely to a gibberellin-dependent release or whether there is also some gibberellin-dependent synthesis (with the exception of the phosphatase which is due to both release and synthesis).

In the aleurone layer of dry barley grains most of the acid phosphatase activity is localized in the inner regions of the cell wall (Ashford and Jacobsen, 1974). During imbibition and incubation of the aleurone layer without gibberellic acid there is increased accumulation of phosphatase activity in the cell wall with no release into the medium. When gibberellic acid is added phosphatase moves into the wall regions that become digested in response to the gibberellic acid. The channels resulting from the gibberellic acid-dependent digestion of the wall presumably provide routes for the release of phosphatase into the medium. It is likely that these channels provide a common path for the release of all the enzyme activities that appear in the medium. The walls appear to have some charged groups because release of α-amylase is very slow at low ionic strengths (Varner and Mense, 1972). It is therefore of physiological importance (with respect to the release of secreted enzymes into the starchy endosperm) that there is also a gibberellin-dependent release or secretion of ions from the aleurone cells (Jones, 1973). This release of Mg^{2+}, Mn^{2+}, Ca^{2+}, K^+, Fe^{2+}, and HPO_4^{2-} (Clutterbuck and Briggs, 1974) is also important to the nutrition of the seedling.

From this brief introduction it should be clear that the aleurone cells are already highly differentiated in the mature seed and poised to respond to gibberellins in a way that will benefit the growing seedling.

Synthesis of Proteins

Although it is now well-established that the response of the aleurone cells to gibberellins includes a synthesis and release of sucrose (Chrispeels, *et al.,* 1973), swelling and dissolution of protein bodies (Jones, 1969), release of organic ions from the phytin globoid of the protein bodies (Jones, 1973), proliferation of rough endoplasmic reticulum (Jones, 1969; Vigil and Ruddat, 1973), release of existing phosphatase, peroxidase, esterase (Ashford and Jacobsen, 1974), and β-glucanase (Jones, 1972) and the synthesis and secretion of α-amylase (Filner and Varner, 1967), protease (Jacobsen and Varner, 1967), β-glucanase and ribonuclease (Bennett and Chrispeels, 1972), it is not generally appreciated to what extent gibberellins interrupt whatever was happening in the aleurone cells and redirect the cells towards an almost exclusive occupation with hydrolase synthesis and secretion. Of twelve proteins released from aleurone layers in response to gibberellic acid, ten become labeled if the aleurone layers are incubated in labeled amino acids (Jacobsen and Knox, 1974). The other two proteins probably represent the release from cell walls of proteins already secreted before the gibberellin treatment. Thus most— perhaps all—of the secreted proteins are synthesized in response to the added gibberellic acid. A dramatic representation of this redirection of protein synthesis is shown in Figures 1 to 4. In both salt-soluble and salt-insoluble proteins the effect of gibberellic acid on the kinds of protein being synthesized is detectable within two hours of the addition of the hormone and marked within two to four hours. After ten hours of treatment with gibberellic acid there is no detectable synthesis of those proteins characteristic of the control tissue. The major labeled protein in the salt-soluble proteins after treatment of the layers with gibberellic acid is α-amylase. The identity of the other labeled proteins is not known. Nor is the identity known of the labeled bands in the salt-insoluble proteins. The addition of abscisic acid along with gibberellic acid prevents the shift in the kinds of proteins being made.

Synthesis of RNA

To what extent does this redirection of protein synthesis require the synthesis of RNA after the addition of gibberellic acid? It is certain that some kind of RNA synthesis is required after the addition of gibberellic acid (Varner, 1964; Chrispeels and Varner, 1967) but it is not certain whether the kinds of RNA synthesized include mRNA for the hydrolases. Carlson (1972) suggests that the mRNA for α-amylase is already present in the imbibed half-seeds before the addition of gibberellic acid and that the added gibberellic acid controls translation. On the other hand, there is a gibberellin-enhanced synthesis of poly(A)-containing RNA (Jacobsen and Zwar, 1974a; Ho and Varner, 1974)—presumed to be mRNA. Because this gibberellin-enhanced synthesis of poly(A)-RNA immediately precedes the rapid synthesis of α-amylase and continues until α-amylase synthesis is no longer susceptible to inhibition by cordycepin (Ho and Varner, 1974), it is reasonable to suppose that the poly(A)-RNA being synthesized is the mRNA for amylase. This can be established only by developing a direct assay for amylase mRNA. This should be possible through the use of the wheat germ cell-free protein synthesizing system (Marcus, et al., 1973).

The increased synthesis of poly(A)-RNA after the addition of GA_3 seems to be the principal—perhaps the only—kind of RNA synthesis required for α-amylase synthesis. Amylase synthesis is not inhibited by concentrations of 5-fluorouracil that severely inhibit rRNA synthesis (Chrispeels and Varner, 1967). And the double labeling technique shows no effect of GA_3 on the synthesis of rRNA, tRNA, and 5sRNA (Jacobsen and Zwar, 1974b). The increase in ribosome numbers [measured by electron microscopy (Jones, 1969) and measured biochemically (Evins, 1971)] and in ribosome turnover (Holmes and Speakman, 1973) apparently do not occur under all conditions (Jacobsen and Zwar, 1974b) and are therefore not required for amylase synthesis.

Role of cAMP

Is cAMP involved in the response of aleurone layers to GA_3? The many reports that the addition to aleurone layers of cAMP by itself or in conjunction with low

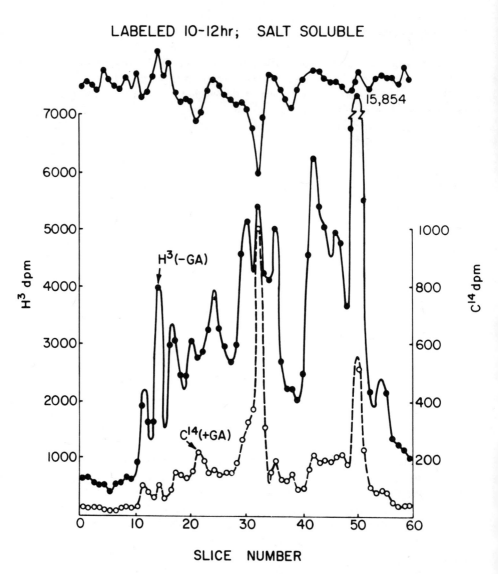

Figure 1.   Labeling of the salt-soluble proteins of control and GA-treated aleurone layers.   The layers were incubated with labeled amino acids for two hours beginning ten hours after the start of the incubation of the layers in buffer, or in buffer plus GA.   The slice numbers refer to the slices obtained from the SDS gels following electrophoresis. The major peak of ^{14}C-activity is due to amylase.   From Varner, Flint and Mitra (1975).

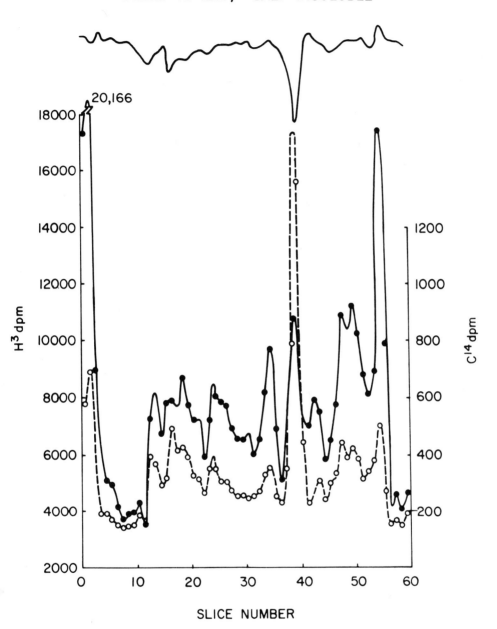

Figure 2. Labeling of the salt-insoluble proteins of control and GA-treated aleurone layers. The layers were incubated with labeled amino acids for two hours beginning ten hours after the start of the incubation of the layers in buffer, or in buffer plus GA. The slice numbers refer to the slices obtained from the SDS gels following electrophoresis. From Varner, Flint and Mitra (1975).

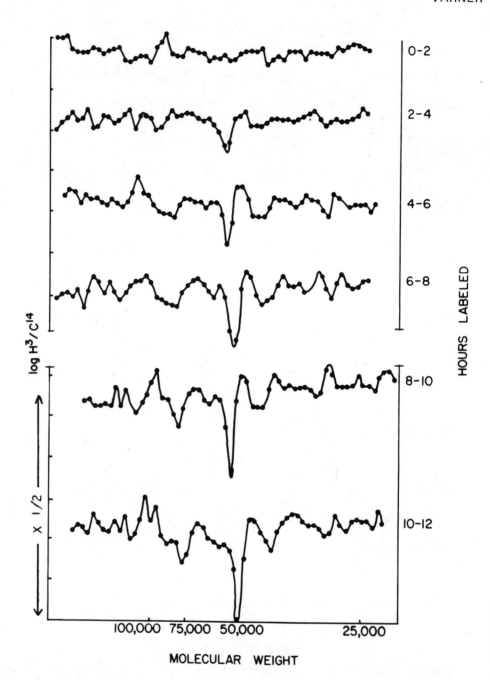

Figure 3. Labeling of the salt-soluble proteins of the control and GA-treated aleurone layers expressed as [3]H (control) to [14]C (GA-treated) ratios at incubation times up to twelve hours. From Varner, Flint and Mitra (1975).

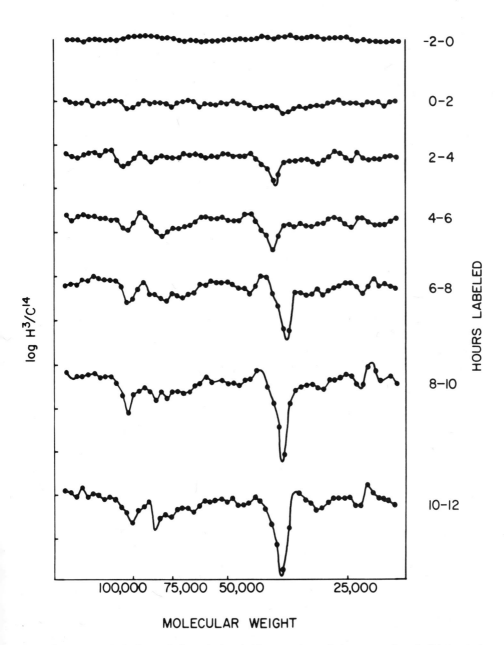

Figure 4. Labeling of the salt-insoluble proteins of the control and GA-treated aleurone layers expressed as ^{3}H (control) to ^{14}C (GA-treated) ratios at incubation times up to twelve hours.

concentrations of GA_3 suggest that cAMP can have a pharmacological effect and that it can cause increased production and/or secretion of amylases or of reducing sugars. There is as yet no convincing evidence that cAMP occurs in barley aleurone layers (or in any plant tissue) at detectable levels (Keates, 1973; Amhrein, 1974a), or that the concentration of cAMP in layers rises to detectable levels after exposure of the layers to GA_3 (Keates, 1973). Furthermore, there is no compelling evidence that there is a specific $3'$, $5'$-phosphodiesterase in higher plants (Lin and Varner, 1972; Amhrein, 1974b).

Osmotic Regulation of Protein Synthesis

The various hydrolases secreted into the endosperm act in concert to digest the polysaccharide and polypeptide reserves. Absorption of the resulting monomers by the scutellum would tend to reduce the total solute concentration of the liquified portion of the endosperm while further hydrolysis of the endosperm would tend to increase the total solute concentration. It appears that there is in fact a physiologically significant osmoregulation of hydrolase synthesis in normally germinating barley seeds (Armstrong and Jones, 1973; Jones and Armstrong, 1971). During the first three days of germination the osmolarity of the starchy endosperm adjacent to the aleurone tissue quickly rises to about 0.5 M and remains more or less constant through the ninth day of germination. Addition of exogenous GA_3 does not change the pattern of osmoticum concentration during germination. Water stress is known to dissociate polysomes to monosomes in many plant cells and addition of 0.6 M polyethylene glycol to barley aleurone tissue that already has a well-developed rough endoplasmic reticulum decreases the ribosome density on the endoplasmic reticulum (Armstrong and Jones, 1973). Such treatment with osmotica in the concentration ranges of 0.2 to 0.6 M decreases α-amylase synthesis (Jones and Armstrong, 1971) and amino acid incorporation into total protein (Chrispeels, 1973) by 25 to 80%.

Secretion of Hydrolases

The physiological effects of the hydrolases that digest the endosperm depend upon their secretion, as well as synthesis, by the aleurone cells. As yet there is no evidence that the process of secretion—i.e. movement of the hydrolases across the plasmalemma— is under the direct control of any hormone or any ion. It has been reported that protease and α-amylase (Gibson and Paleg, 1972) inside the cell are localized in lysosomal-like vesicles and the conclusion drawn that these hydrolases—and perhaps others—are secreted without ever being outside the endoplasmic reticulum-lysosomal system of membranes. On the other hand a radioautographic localization of the proteins synthesized in the presence of GA_3 indicated that nearly all of the protein labeled with [3H]-amino acids in a ten minute pulse was secreted without becoming associated with any vesicle or particle (Chen and Jones, 1974). It appears that the mode of secretion must be established before possible controls of secretion can be considered.

Receptor Site or GA$_3$

There is now a detailed description of the later effects of GA$_3$ on the functioning of the aleurone tissue. But of course there are early effects so far not seen. The GA$_3$ must bind to some kind of receptor site and the GA$_3$-receptor pair must then initiate one or more primary events which lead to the easily observed changes in function of the aleurone cells. The formation of the GA$_3$-receptor complex cannot involve translation or transcription. Initiation of the primary event(s) may involve translation and/or transcription.

Phosphorylcholine Transferases

Increases in phosphorylcholine cytidyl transferase and in phosphorylcholine glyceride transferase (measured in cell-free preparations) apparently begin within minutes

TABLE I

THE EFFECT OF GA$_3$ AND AMINO ACID ANALOGS ON PHOSPHORYLCHOLINE GLYCERIDE TRANSFERASE ACTIVITY IN BARLEY ALEURONE*

Treatment during incubation	% of control	Standard*** deviation
Control	100**	
10^{-6} M GA$_3$	161	$\pm$38.2
10^{-6} M GA$_3$ and 7x10^{-3} M amino acid analogs	167	$\pm$40.4

*Enzyme prepared from aleurone layers after 4 hr incubation of half seeds.

**The controls were equal to 1229, 1022 and 1435 pmoles lecithin formed/hr/100 layers in three different experiments respectively, each one with two replicates.

***Because of large differences in the absolute pmoles of lecithin formed in different experiments the control in each experiment has been taken arbitrarily as 100% and the treatments effect was compared on a percent basis in each experiment separately, so that the standard deviations for the treatments would be calculated. From Ben-Tal and Varner (1974).

of the addition of GA_3 to aleurone layers (Johnson and Kende, 1971). The increase in phosphorylcholine glyceride transferase apparently does not require protein synthesis (Ben-Tal and Varner, 1974). The requirement for translation for the GA_3-dependent increase in phosphorylcholine transferase was tested by incubating aleurone layers in the presence of seven amino acid analogs known to be incorporated into proteins. The rationale was that enzymes synthesized in the presence of the amino acid analogs would be nonfunctional due to extensive substitution of the analogs for the normal amino acids, whereas enzyme activities that increase due to activation should not be affected by the presence of the analogs. The amino acid analogs had no effect on the GA_3-dependent increase in phosphorylcholine glyceride transferase activity (Table I) but did inhibit the nitrate-induced formation of nitrate reductase (Ben-Tal and Varner, 1974). Thus the analogs did enter the aleurone cells and one can conclude that protein synthesis is not involved in the early GA_3-dependent increase in phosphorylcholine glyceride transferase. If protein synthesis is not required for this early response of the aleurone cell to GA_3, it is not likely that RNA synthesis would be required. Concentrations of cordycepin sufficient to prevent nitrate-induced formation of nitrate reductase (Ben-Tal and Varner, 1974) had no effect on the GA_3-dependent increase in phosphorylcholine glyceride transferase (Table II).

TABLE II

THE EFFECT OF GA_3 AND CORDYCEPIN ON PHOSPHORYLCHOLINE GLYCERIDE TRANSFERASE ACTIVITY IN BARLEY ALEURONE*

Treatment during incubation	% of control	Standard*** deviation
Control	100**	
10^{-6} M GA_3	164	±13.4
10^{-6} M GA_3 and 10^{-3} M cordycepin	147	±18.9

*Enzyme prepared from aleurone layers after 4 hr incubatin of half seeds.

**The controls were equal to 1162, 2246 and 1175 pmoles lecithin formed/hr/100 layers in three different experiments respectively, each one with two replicates.

***Because of large differences in the absolute pmoles of lecithin formed in different experiments, the control in each experiment has been taken arbitrarily as 100% and the treatments effect was compared on a percent basis in each experiment separately, so that the standard deviation for the treatments would be calculated. From Ben-Tal and Varner (1974).

TABLE III

OBSERVED AND EXPECTED PHOSPHORYLCHOLINE GLYCERIDE TRANSFERASE ACTIVITIES IN MIXTURES OF ENZYME PREPARATIONS FROM GA₃ AND CONTROL ALEURONE LAYERS*

Treatment	pmoles lecithin formed/hr/25 layers		pmoles lecithin formed/hr/mg protein	
	Observed	Expected	Observed	Expected
Control	742		785	
10^{-6} M GA$_3$	1384		1362	
Mixture	1429	1063	1557	1169

A mixture was prepared from equal volumes of crude homogenates ground separately from control tissue and from GA$_3$-treated tissue. The expected value is calculated by adding the activities of the control and the GA$_3$ homogenates and dividing by 2.

*Enzyme prepared from aleurone layers after 4 hr incubation of half seeds in the solutions shown. From Ben-Tal and Varner (1974).

If, as the data suggest, some kind of activation accounts for the early rise in phosphorylcholine glyceride transferase activity following GA$_3$ treatment, one might expect to see some evidence for such activity upon mixing homogenates of GA$_3$-treated tissue with homogenates of control tissue. Such an effect is observed (Tables III and IV) and this activation effect is not observed if abscisic acid is present during the time or treatment of the tissue with GA$_3$ (Table IV).

Is this activation one of the primary events of GA$_3$ action? Phosphorylcholine glyceride transferase activity is inhibited by Ca^{2+}. Therefore the observed changes in activity could result from changes in Ca^{2+} levels or changes in the levels of Ca^{2+} complexing agents—such as citrate or phytic acid. Phosphorylcholine glyceride transferase activity is a membrane-bound enzyme and its activity would likely be influenced by the concentration of surfactants—such as lysolecithin—that might well be produced during the breakdown of storage phospholipids. These questions remain to be answered but it appears that the activation of phosphorylcholine glyceride transferase, if not a primary event, may provide a clue to a primary event in the response of the aleurone tissue to added gibberellic acid.

TABLE IV

THE REVERSAL EFFECT OF ABSCISIC ACID (ABA)
ON THE GA_3-DEPENDENT STIMULATION OF THE ENZYME ACTIVITY

Treatment during incubation	pmoles lecithin formed per					
	25 aleurone layers			mg protein		
	observed	expected	% increase	observed	expected	% increase
1) Control	926			494		
2) 10^{-6} M GA_3	1523			707		
3) 10^{-5} M ABA	898			501		
4) 10^{-6} M GA_3 + 10^{-5} M ABA	852			459		
A mixture of homogenates from treatments (1) and (2)	1713	1227	40	781	600	30
A mixture of homogenates from treatments (2) and (3)	1685	1213	39	803	604	33
A mixture of homogenates from treatments (1) and (4)	954	889	7	477	476	0
A mixture of homogenates from treatments (3) and (4)	963	875	10	507	480	5

Enzyme prepared from aleurone layers after 4 hr incubation of half seed in the solutions shown. Whenever GA_3 or ABA were used they were added to the incubating media at zero time. Mixtures were made out of equal volumes of crude homogenates prepared from the tissue incubated in the specific solution. From Ben-Tal and Varner (1974).

Acknowledgments

This work was supported in part by the National Science Foundation (GB-33944).

REFERENCES

Amhrein, N. (1974a). *Planta 118,* 241.

Amhrein, N. (1974b). *Z. Pflanzenphysiol. 72,* 249.

Armstrong, J.E. and Jones, R.L. (1973). *Cell. Biol. 59,* 444.

Ashford, A.E. and Jacobsen, J.V. (1974). *Planta. 120,* 81.

Bennett, P.A. and Chrispeels, M. (1972). *Plant Physiol. 49,* 445.

Ben-Tal, Y. and Varner, J.E. (1974). *Plant Physiol. 54,* 813.

Carlson, P.S. (1972). *Nature 237,* 39.

Chen, R. and Jones, R.L. (1974). *Planta 119,* 207.

Chen, S. and Park, W. (1973). *Plant Physiol. 52,* 174.

Chrispeels, M.J. (1973). *Biochem. Biophys. Res. Comm. 53,* 99.

Chrispeels, M.J., Tenner, A.J. and Johnson, K.D. (1973). *Planta 113,* 35.

Chrispeels, M.J. and Varner, J.E. (1967). *Plant Physiol. 42,* 1008.

Clutterbuck, V.J. and Briggs, D.E. (1974). *Phytochem. 13,* 45.

Evins, W.H. (1971). *Biochem. 10,* 4295.

Filner, P. and Varner, J.E. (1967). *Proc. Nat. Acad. Sci. USA 58,* 1520.

Gibson, R.A. and Paleg, L.G. (1972). *Biochem. J. 128,* 367.

Ho, D. and Varner, J.E. (1974). *Proc. Nat. Acad. Sci. USA. 71,* 4183

Holmes, P.L. and Speakman, P.T. (1973). *Nature 242,* 190.

Jacobsen, J.V. and Knox, R.B. (1974). *Planta 115,* 193.

Jacobsen, J.V. and Varner, J.E. (1967). *Plant Physiol. 42,* 1596.

Jacobsen, J.V. and Zwar, J.A. (1974a). *Proc. Nat. Acad. Sci. USA 71,* 3290.

Jacobsen, J.V. and Zwar, J.A. (1974b). *Austral. Jour. Plant Physiol.* (In press.)

Jones, R.L. (1969). *Planta 88,* 73.

Jones, R.L. (1972). *Planta 103,* 95.

Jones, R.L. (1973). *Plant Physiol. 52,* 303.

Jones, R.L. and Armstrong, J.E. (1971). *Plant Physiol. 48,* 137.

Keates, R.A.B. (1973). *Nature 244,* 355.

Lin, P.P. and Varner, J.E. (1972). *Biochim. Biophys. Acta 276,* 454.

Marcus, A., Efron, D. and Weeks, L.P. (1973). *Methods in Enzymology,* ed. by Academic Press.

Varner, J.E. (1964). *Plant Physiol. 39,* 413.

Varner, J.E., Flint, D. and Mitra, R. (1975). *Workshop on Genetic Improvement of Seed Proteins.* National Academy of Sciences/National Research Council. (In press.)

Varner, J.E. and Mense, R. (1972). *Plant Physiol. 49,* 187.

Vigil, L. and Ruddat, M. (1973). *Plant Physiol. 51,* 549.

Yomo, H. and Varner, J.E. (1971) in *Current Topics in Developmental Biology,* eds. Moscana, A. and Monroy, A., Vol. 6 (New York: Academic Press), pp. 111-144.

OVALBUMIN mRNA AND OVALBUMIN DNA AND THE MOLECULAR BIOLOGY OF STEROID HORMONE ACTION

*Robert T. Schimke, David J. Shapiro,[1] and
G. Stanley McKnight[2]

The Department of Biological Sciences
Stanford University
Stanford, California 94305

Introduction

Certain steroid hormones, including estrogens, progesterone, and testosterone have profound effects on the development and function of the hen oviduct (Oka and Schimke, 1969a, b; O'Malley et al., 1969; Palmiter and Wrenn, 1971; Cox and Sauermein, 1970). Our attention has focused on hormonal regulation of ovalbumin synthesis, since this single polypeptide constitutes 50-60% of the protein synthesized in the fully differentiated oviduct, and its synthesis is under control by estrogens and progesterone. These features have allowed an analysis of the regulation at the molecular level, most specifically characterization and quantitation of the elements involved in specific protein synthesis, including polysomes, mRNA, and genes, in order to determine which of a myriad of potentially rate limiting steps is regulated by the hormones. We describe herein some of our more recent studies on the regulation of ovalbumin synthesis, including characterization of ovalbumin mRNA, and the use of nucleic acid hybridization techniques to analyze the mechanism of steroid hormone action.

---

*Symposium participant.

[1]Helen Hay Whitney Foundation Fellow, current address: Department of Biochemistry, University of Illinois, Champaign-Urbana, Illinois.

[2]Current address: Faculte de Medecine, Institut de Chimie Biologique, Strasbourg, France.

Hormonal Regulation of Oviduct Development and Ovalbumin Synthesis

Figure 1 summarizes the effects of estrogens and progesterone on oviduct development and function. Estrogen administration to immature chicks (denoted primary stimulation) results in cytodifferentiation of tubular gland cells, which synthesize the major
egg white proteins, including ovalbumin, conalbumin, and lysozyme (Oka and Schimke,
1969b; Palmiter and Gutman, 1972). When estrogen administration is stopped (denoted
withdrawal), the cells persist, but ovalbumin synthesis ceases. The reinitiation of ovalbumin synthesis in tubular gland cells (denoted secondary stimulation) can occur on
pre-existing ribosomes (Palmiter *et al.*, 1970) and can be produced by the administration
of either estrogen or progesterone. Concomitant administration of progesterone with
estrogen during primary stimulation, in contrast, prevents typical cytodifferentiation
produced by estrogen alone (Oka and Schimke, 1969b); rather, an abortive synthesis of
ovalbumin occurs in the surface epithelial cells, but such cells fail to undergo typical
changes inculding proliferation and formation of tubular gland cells (Palmiter and Wrenn,
1971).

Figure 2 shows the percentage of protein synthesis that constitutes ovalbumin
during different stages of estradiol administration and withdrawal. Ovalbumin synthesis
is measured by incubating small fragments of oviduct tissue for 30 minutes in a salt
medium containing labeled amino acids and subsequently precipitating ovalbumin from
labeled supernatant proteins with specific antibody. In the immature chick, no ovalbumin
synthesis can be detected (Palmiter and Wrenn, 1971). Capacity for ovalbumin synthesis

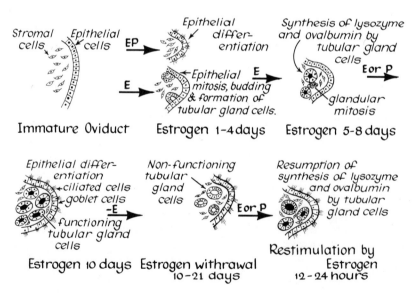

Figure 1. Hormone interactions in development and function of tubular gland cells
of chick oviduct. E, estrogen, P, Progesterone.

increases such that by ten days of daily estrogen administration, as much as 50% of the protein synthesized is ovalbumin. Withdrawal of hormone treatment results in a rapid decrease in ovalbumin synthesis to undetectable amounts. During secondary stimulation, there is a rapid increase in ovalbumin synthesis. The lag in ovalbumin synthesis during primary stimulation results from the necessity for cytodifferentiation of tubular gland cells, whereas the rapid increase during secondary stimulation occurs in previously existing tubular gland cells (Oka and Schimke, 1969b).

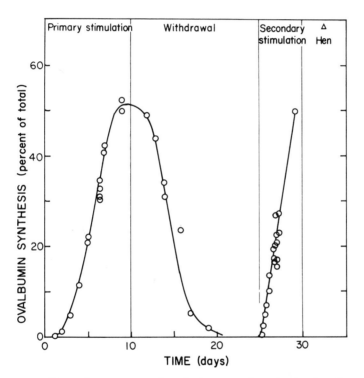

Figure 2. Effect of estrogen on the relative rate of synthesis of ovalbumin in chick oviduct during primary stimulation, withdrawal, and secondary stimulation. Immature chicks, four days old, were injected intramuscularly with 1 mg estradiol benzoate daily (primary stimulation) and after ten days without estrogen administration (withdrawal) administration was resumed (secondary stimulation). Fragments of oviduct were then incubated in Hank's salt solution for one hour with tritium-labeled amino acids (10 μCi/ml). Following homogenization and centrifugation at 100,000g for one hour, ovalbumin was precipitated from the supernatant using a specific antibody. Results are presented as percentage of total acid-precipitable radioactivity in supernatant that is precipitated immunologically with anti-ovalbumin antibody. Details are given in Palmiter and Schimke (1971).

Theoretically, the hormones can stimulate specific protein synthesis by a variety of mechanisms, including amplification of genes, stimulation of mRNA transcription, enhanced mRNA transport from nucleus to cytoplasm, activation of cytoplasmic mRNA, etc. Critical to an analysis of potential regulatory steps is the ability to purify and characterize ovalbumin mRNA and to develop methods for its identification and quantitation, as well as methods for analysis of the ovalbumin gene.

Isolation and Characterization of Ovalbumin mRNA

For isolation of ovalbumin mRNA, we have employed the rabbit reticulocyte lysate system for its identification (Palmiter and Schimke, 1973). Increasing amounts of RNA preparations from oviduct tissue, when added to the lysate system, produce a linear increase in radioactivity that is precipitable with a specific ovalbumin antibody (Palmiter and Schimke, 1973). We have characterized the product that is immunoprecipitated as ovalbumin by the following criteria: (a) it is precipitated only with a specific antibody; (b) it is the same size as authentic ovalbumin; (c) tryptic peptides co-chromatograph with those of authentic ovalbumin (Rhoads et al., 1971).

For isolation of the ovalbumin mRNA, we have developed a technique based on the ability of antibody to ovalbumin to bind specifically to ovalbumin nascent chains (Palacios et al., 1972; Palmiter et al., 1972; Palacios et al., 1973) which can thereby be specifically precipitated by several immunologic techniques, including reaction of the polysome-antibody complex with a matrix of glutaraldehyde-treated ovalbumin (Palacios et al., 1973), or with anti-antibody (Shapiro et al., 1974). We are currently employing the indirect immunoprecipitation technique involving reaction of polysomes with antibody, followed by use of an anti-antibody (generally goat anti-rabbit), and have found this technique applicable to the isolation of serum albumin (Payvar and Schimke, unpublished results) and conalbumin polysomes (Schimke et al., 1973), as well as the ovalbumin synthesizing polysomes reported herein. We believe that this is the best general method for isolation of polysomes, since it does not depend on unusual physical characteristics of a mRNA, and can be used to isolate one size of specific mRNA from among others of the same size class, a separation that is not possible when mRNA is isolated only on the basis of size alone (i.e. by sucrose gradients or various types of gels). Details of the methods and proof of specificity are given in the cited references and summarized by Schimke et al. (1973).

The fraction of RNA isolated by immunologic techniques contains specific ovalbumin mRNA as well as a large amount of accompanying ribosomal RNA. Isolation of ovalbumin mRNA is best achieved in our hands (Shapiro and Schimke, 1975) by poly (U)-sepharose chromatography (Firtel and Lodish, 1973) and is based on the fact that the ovalbumin mRNA, like most mRNAs isolated from eukaryotes contains poly(A) sequences at the 3' OH end (Brawerman et al., 1972). Employing indirect immunoprecipitation to separate ovalbumin synthesizing polysomes from total polysomes and poly (U)-sepharose to separate the ovalbumin mRNA from ribosomal RNA, ovalbumin mRNA has been purified 90 to 95-fold from the starting polysomal preparation and with 40% yield based on an *in vitro* translation assay (Shapiro and Schimke, 1975). A similar fold purification is found by methods involving the rate of hybridization of complementary DNA (cDNA) to purified mRNA and to total polysomal RNA (97-fold). This latter method is independent of the "translatability" of the mRNA and since both methods indicate essentially the same fold purification, we conclude that our purification method does not result in progressive inactivation of translatable mRNA.

Isolation of ovalbumin mRNA by indirect immunoprecipitation and poly (U)-sephrose chromatography does not involve purification on the basis of size, since no sucrose gradients are employed. Figure 3 shows optical scans of the purified ovalbumin mRNA as sedimented on sucrose gradients (Fig. 3a) and electrophoresed on acrylamide gels (Fig. 3b). The ovalbumin mRNA sediments as a single symmetrical peak with an S value of approximately 16. The purified mRNA migrates on acrylamide gels as a single sharp peak with an S value of approximately 20 S. This anomalous behavior questions the correct size of mRNA and the reasons for different estimates of size dependent on the method used.

Since the rate of sedimentation of macromolecules is dependent on their secondary structure, we estimated the size of the ovalbumin mRNA under denaturing conditions in sucrose gradients containing dimethylsulfoxide (Strauss *et al.*, 1963). Figure 4 shows that under denaturing conditions, ovalbumin mRNA co-sediments with 18 S ribosomal RNA from the chick, an RNA that is estimated to contain 2200 nucleotides.

The size of a nucleic acid molecule can also be estimated by determining the rate of hybridization of the ovalbumin mRNA to DNA that is complementary to the mRNA. The synthesis of such complementary DNA (cDNA) is described later. Figure 5 shows the kinetics of hybridization of ovalbumin cDNA to both the highly purified mRNA, and to total polysomal RNA. The two curves are similar, both with respect to the extent of hybridization under the conditions employed, as well as the shape of the curve. When such data is plotted as double reciprocal plots (Birnstiel *et al.*, 1972), the data fits a single, straight line, indicating that the cDNA is hybridizing to a single molecular species of RNA. The acceleration in rate of hybridization is 92-fold (comparing Crt ½ values) and therefore ovalbumin mRNA sequences have been enriched by that amount. The size of the ovalbumin mRNA can now be calculated by comparison to the Crt ½ and known complexity of poliovirus RNA using the simple expression:

$$\frac{\text{Crt } \frac{1}{2} \text{ Ov mRNA}}{\text{Crt } \frac{1}{2} \text{ polio mRNA}} = \frac{\text{complexity of Ov mRNA}}{\text{complexity of polio mRNA}}$$

Polio virus mRNA has a complexity of 7,500 nucleotides (Baltimore *et al.*, 1969), and a Crt ½ under our hybridization conditions of 4.7×10^{-3} M sec/L. Therefore, the complexity of ovalbumin mRNA is 2,150 nucleotides, which corresponds to a molecular weight of 700,000. This is in good agreement with the value of 2,200 nucleotides obtained from the dimethylsulfoxide-sucrose gradients.

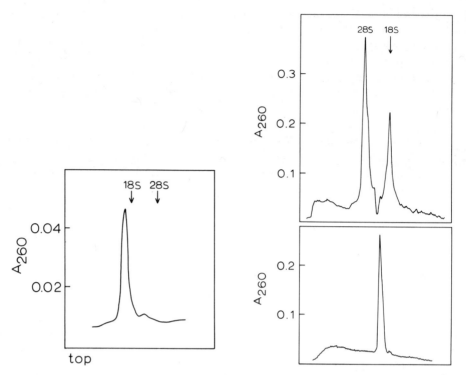

Figure 3. (a) Sedimentation profile of ovalbumin mRNA on sucrose gradients. Purified ovalbumin mRNA (1.8 μg) dissolved in 50 μl of 25 mM tris pH 7.1, containing 1% sodium dodecyl sulface, 2 mM EDTA was layered over a linear 5-20% sucrose gradient and sedimented for 4½ hr at 225,000g. Polysomal RNA markers were run in separate tubes.

(b) Acrylamide gel electrophoresis of purified ovalbumin mRNA. 8 μg of oviduct polysomal RNA and 4 μg of purified ovalbumin mRNA were dissolved in 25 μl of electrophoresis buffer containing 10% glycerol and bromophenol blue. Pre-electrophoresis of the 2.5% acrylamide 14 x 0.6 cm gels was for 45 min. The gel was scanned after electrophoresis in a Gilford Spectophotometer with a linear transit. Background A_{260} peaks near the bottom of the gel and trails off. The background is not due to RNA since it is found on blank gels following electrophoresis. After scanning, the gel was sliced, RNA eluted and assayed for ovalbumin mRNA activity, which was exclusively localized under the major optical density peak. Details are given in Shapiro and Schimke (1975).

The minimum number of nucleotides required for ovalbumin synthesis is 1,161 nucleotides (387 amino acids X 3), or approximately 1000 nucleotides less than actually are present in ovalbumin mRNA. Ovalbumin mRNA contains a poly(A)-sequence, which we estimate to contain on the average a sequence of 40 nucleotides in length (Shapiro and Schimke, 1975). Since there is no indication that ovalbumin is synthesized as a large molecular weight precursor (Palmiter and Schimke, 1973) and Figure 6, ovalbumin mRNA appears to contain approximately 1000 nucleotides which are not translated. Current evidence indicates that other eukaryote mRNAs appear to contain a similar proportion of untranslated sequences (Breindl and Galliwitz, 1973; Mach et al., 1973; Berns et al., 1973). The nature and possible function of these sequences in translation and transcription are currently under investigation.

The proof of the purity of ovalbumin mRNA is based on a number of lines of evidence. First is the specificity of the immunoprecipitation techniques, which allow for at most 0.4% non-specific trapping of polysomes (Shapiro et al., 1974). Second is the single peak of RNA evidenced by sucrose gradient centrifugation and acrylamide gel electrophoresis (Fig. 3). Third is our estimate that a pure preparation of ovalbumin mRNA should theoretically be 90-fold, starting with total polysomes (from total oviduct RNA, this value would be 180-fold, since the polysomes as we isolate them constitute 50%

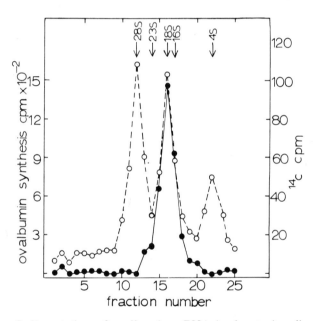

Figure 4. Sedimentation of ovalbumin mRNA in denaturing dimethylsulfoxide gradients. ^{3}H-labeled hen oviduct RNA, (1,200 cpm/µg) and 0.2 µg of purified ovalbumin mRNA were dissolved in a mixture of water, dimethylformamide and dimethylsulfoxide (Firtel and Lodish, 1973; Strauss et al., 1968) heated to 60° for 5 min and cooled rapidly. Sedimentation on 99% dimethylsulfoxide 0-8% sucrose gradients for 18 hr at 30° at 270,000g. Fractions were collected, diluted with 2 volumes of water containing 0.2 M sodium acetate and RNA precipitated with ethanol. The RNA was redissolved and an aliquot was counted to locate the labeled chick RNA (o---o). The remainder of each fraction was assayed for ovalbumin mRNA activity (●---●). E. coli 16S and 23S rRNA markers were sedimented in parallel gradients. The E. coli optical density profiles are omitted for simplicity. Details are given in Shapiro and Schimke (1975).

of total RNA (Shapiro and Schimke). The last evidence involves the use of the wheat germ cell-free protein synthesizing system, which is totally dependent on the addition of exogeneous mRNA (Marcus *et al.*, 1973). If a mRNA preparation contained contaminating mRNAs, these would also be translated and appear as protein peaks on acrylamide gels different from that of ovalbumin (assuming that the proteins are of different molecular weight). Figure 6 shows a sodium dodecyl sulfate acrylamide gel of the protein products synthesized in a wheat germ system with added 95-fold purified ovalbumin mRNA and total oviduct polysomal RNA (containing multiple mRNAs). The upper panel shows the pattern of incorporation with total mRNAs showing various sized peaks of radioactivity. The lower panel shows the results with ovalbumin mRNA. In the latter, there is a single peak of radioactivity migrating with authentic ovalbumin. That the counts are, indeed, only in ovalbumin is further proven by the fact that radioactivity immunoprecipitated with ovalbumin antibody represents essentially all of the radioactivity incorporated in the presence of ovalbumin mRNA, and electrophoresis of the immunoprecipitated radioactivity gives the same result as shown (lower panel of Fig. 6) for total incorporation. Assuming that various mRNAs are equally well utilized in the wheat germ system, we conclude by this translation assay, that the ovalbumin mRNA is essentially homogeneous.

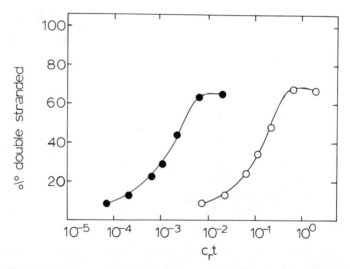

Figure 5. Kinetics of reassociation of purified ovalbumin mRNA and oviduct polysomal RNA with complementary DNA. Purified ovalbumin mRNA (.00048 A_{260}/ 10 μl hybridization) (●---●), or total oviduct polysomal RNA (.048 A_{260}/10 μl hybridization) (○---○), were mixed in 10 μl with 300 cpm single stranded complementary DNA (10^4 cpm/ng) prepared against purified ovalbumin mRNA. Hybridization was at 68° for 0-24 hr, and hybrids were scored. The background obtained by counting a blank filter is subtracted from all values. Continuing the hybridization for an additional 4 days did not appreciably increase the percent hybridized. Details are given in Shapiro and Schimke (1975).

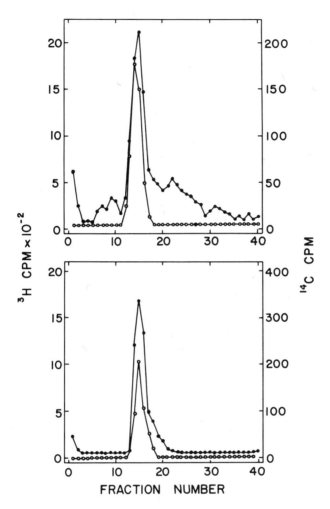

Figure 6. Sodium dodecyl sulfate acrylamide gel electrophoresis of labeled proteins from a wheat germ protein synthesizing system incubated with total oviduct polysomal RNA or purified ovalbumin mRNA. Wheat germ (General Mills) was ground in a mortar and pestle without prior fractionation and lysate was prepared and freed of endogenous amino acids essentially as described by Marcus (1973). 5 μg of total oviduct polysomal RNA (top panel) or 0.05 μg of purified ovalbumin mRNA (bottom panel) were incubated in a total volume of 75 μl at 25° for 60 min. Marker [14C] ovalbumin (o---o) and the ^{3}H-synthesized proteins (●---●) were subjected to electrophoresis in 10% acrylamide gels as described by Shapiro *et al.* (1974).

Synthesis of Complementary DNA

We have used the RNA-dependent DNA polymerase of Rous Sarcoma Virus (RSV) (Faras *et al.*, 1972) to synthesize a nucleic acid sequence complementary to ovalbumin mRNA. This enzyme uses as a template single-stranded RNA with a short primer region of double-stranded nucleic acid at the 3' end. By addition of oligo (dT) to the purified ovalbumin mRNA preparation, the mRNA molecules containing the putative poly(A) sequence are converted to a double stranded primer for the RNA dependent DNA polymerase reaction.

As shown in Table I, an ovalbumin mRNA fraction that has been specifically immunoprecipitated and selectively adsorbed on Millipore filters is active as a template for RNA-dependent DNA polymerase in a system that contains RSV reverse transcriptase and oligo (T) (Sullivan *et al.*, 1973). That the reaction is dependent on RNA is shown by the fact that RNase treatment of the RNA completely abolishes incorporation of deoxynucleoside triphosphate into an insoluble form. Likewise, the reaction is essentially totally dependent on oligo (T) addition.

That the ovalbumin cDNA is complementary only to ovalbumin mRNA sequences is demonstrated in Figure 7, which shows hybridization of the cDNA to pure ovalbumin

TABLE I

TEMPLATE ACTIVITY OF RNAs

RNA added to polymerase system	Acid precipitable ^{3}H cpm/50 μl reaction
+ ovalbumin messenger fraction (1.3 μg/ml)	79,000
+ ovalbumin messenger fraction + RNase	250
polymerase system alone (no added RNA)	950
polymerase system alone + RNase	200
+ ovalbumin messenger fraction, no oligo (dT)	900

In those samples described as "+RNase" the added RNA, if any, was preincubated thirty min at 37° with 100 μg/ml boiled pancreatic RNase before this mixture was added to the polymerase reaction (see Sullivan *et al.* (1973) for details).

mRNA, to total polysomal RNA, and to total oviduct RNA (McKnight *et al.,* unpublished results). The results are presented as a double reciprocal transformation (Bishop, 1972) of the percent protection of the cDNA hybridized to RNA as detected by *Aspergilis* single stranded nuclease (S_1). Although the rate of hybridization is different for the three types of RNA (evidenced by the different slopes), the percent hybridization at infinite RNA time concentration (Crt) is the same for all three samples. If there was a fraction of cDNA that hybridized with RNA species present in polysomal or total RNA preparations, that were not present in the pure ovalbumin mRNA, we would have observed a greater extent of hybridization. That we observe the same extent of hybridization indicates that there

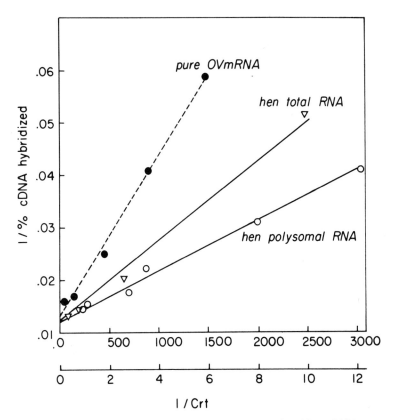

Figure 7. Double reciprocal plot of hybridizations of oviduct RNA to ovalbumin cDNA. Pure ovalbumin mRNA (●---●) at 3.8×10^{-3} A_{260}/ml was hybridized to approximately 40 pg of cDNA in 10 μl. Hen total RNA (△---△) at 4.38×10^{-2} A_{260}/ml and hen polysomal RNA (○---○) at $3.52 \times 10^2 A_{260}$/ml were hybridized to 20 pg of cDNA in 10 μl volumes. The upper scale on the abcissa is for pure ovalbumin mRNA and the lower scale is for both total and polysomal hen RNA. Details are given in McKnight *et al.* (submitted for publication).

are no species of RNA in the total RNA preparation other than ovalbumin mRNA that react with the cDNA. Hence the cDNA is specific for ovalbumin mRNA.

Three lines of evidence indicate that the cDNA is not the full length of the ovalbumin mRNA. We estimate (see above) that the ovalbumin mRNA contains approximately 2200 nucleotides, only 40 of which are poly(A) sequences at the 3' OH end. Sedimentation of the cDNA synthesized in the presence of actinomycin D and hence single stranded (Sullivan *et al.*, 1973) on a sucrose gradient indicates that the mean S value is approximately 7-9S, suggesting a nucleotide length of approximately 400. We have made an additional estimate of the length of double stranded cDNA by a determination of the complexity of the reassociation reaction of double stranded cDNA (*i.e.* synthesized in the absence of actinomycin D) and estimate that the nucleotide length is approximately 200-400 nucleotides (Sullivan *et al.*, 1973). Lastly, we have recently hybridized increasing amounts of pure ovalbumin mRNA in the presence of excess cDNA under conditions where the hybridization is complete and find that approximately 24% of the mRNA forms a mRNA-cDNA hybrid, indicating that approximately 400-500 nucleotides have been protected (McKnight *et al.*, unpublished results). Taken together, these three types of results indicate that the reverse transcriptase initiates at the 3' OH end of the mRNA containing the poly(A) sequence, and synthesizes a cDNA that is approximately 400-500 nucleotides in length. The reasons that a complete copy of the ovalbumin mRNA is not synthesized (*i.e.* 2200 nucleotides) is not known.

The approximately 400 nucleotide-long segment is sufficient to obtain accurate and stable nucleic acid hybrids, and hence can be used for determining the number of DNA and RNA sequences in the oviduct under different hormonal states.

Analysis of the Number of Ovalbumin Genes

The DNA probe can be used to determine the number of ovalbumin genes and thereby answer the question of whether there is differential gene amplification to account for the large amount of ovalbumin synthesis in the oviduct. To determine the number of genes, the experiment shown in Figure 8 was performed (Sullivan *et al.*, 1973). A large amount of unlabeled chicken liver DNA or oviduct DNA was melted and allowed to reanneal in the presence of trace amounts of ^{14}C-labeled chick fibroblast DNA that had previously been fractionated on hydroxylapatite columns to remove all but the "unique" sequence DNA, as well as trace amounts of [3H] ovalbumin specific single-stranded probe. The rate of reassociation of the labeled DNA will depend on the concentration of complementary sequences in the unlabeled DNA, since these are in excess. Figure 8 clearly indicates that both ovalbumin specific and "unique sequence" DNA reassociate at the same rate.

Since the cDNA is complementary to at most the 400-500 nucleotides adjacent to the 3' OH end of the mRNA, and since we currently have no knowledge of whether these sequences among the 2200 present in ovalbumin mRNA are involved in the coding for the

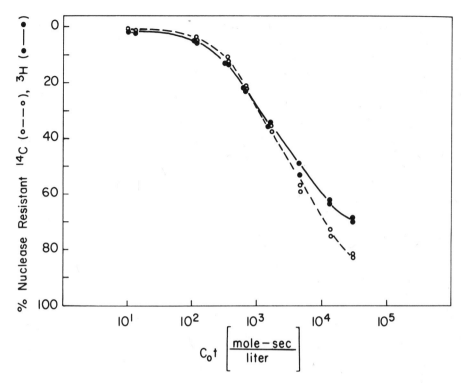

Figure 8. Determination of the absolute copy number of ovalbumin sequences per chicken oviduct genome. Chicken oviduct DNA at 10 mg/mg (prepared by the technique of Varmus *et al.,* 1971), ^{14}C-unique-sequence DNA at 30 µg/ml (o--o) and ^{3}H-single stranded probe DNA at about 1 ng/ml (●--●) were mixed together, then melted, and reannealed. At different times aliquots were taken and assayed for S_1 nuclease resistance. The data is plotted relative to the Cot of the unlabeled liver DNA. See Sullivan *et al.* (1973) for details.

structure of ovalbumin, it remains conceivable that the ovalbumin gene is, indeed, re-peated in the genome, and that the 500 nucleotide unit recognized by the cDNA may be some type of spacer region in the ovalbumin mRNA which has undergone random mutation, and hence may be scored as unique sequence DNA. This possibility has been ruled out by direct hybridizations of mRNA to chick DNA where the pure mRNA has been labeled with [125I] ovalbumin mRNA (Shapiro and Schimke, unpublished results). The results (Fig. 9) show that the labeled mRNA also hybridizes with unique sequence DNA. From the data, it appears that the vast majority of the [125I] ovalbumin mRNA hybridizes at Cot values characteristic of unique sequence DNA. Figure 9 also shows that a portion of the iodinated ovalbumin mRNA hybridizes to DNA reiterated approximately 50-100 times in the chick genome. This constitutes approximately 12-15 percent of the radioactivity and would correspond to a sequence in the mRNA of approximately 250 nucleotides. Since this sequence is not sufficiently long to correspond to that required to code for ovalbumin (1161 nucleotides), we conclude that the DNA coding for ovalbumin is truly represented only once per haploid genome. Under the same conditions as employed in Figure 9, calf thymus DNA does not hybridize to ovalbumin mRNA. In addition, the hybridization of iodinated mRNA to repetitive DNA is not completed by addition of unlabeled ribosomal RNA, and hence the hybridization does not represent contamination with this RNA species. We are currently considering the possibilities that such sequences are related to membrane binding properties of the mRNA, or are common to various mRNAs responsive to steroid hormones.

Identification and Quantitation of mRNA Sequences Using cDNA

cDNA can be employed in RNA-cDNA hybridization reactions to determine the

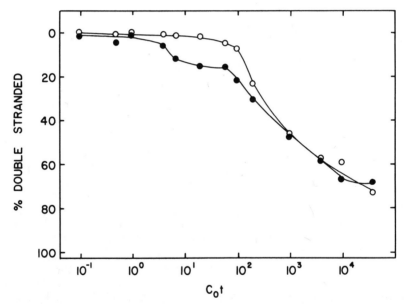

Figure 9. Determination of the number of ovalbumin gene copies with iodinated ovalbumin mRNA. Purified ovalbumin mRNA was iodinated by a modification of the method of Comerford (1971). The RNA was heated to 68° for 15 min twice to remove non-covalently bound iodine and then purified on a 5-20% sucrose gradient containing sodium dodecyl sulfate to isolate the full-length (16S) mRNA molecules. Iodinated ovalbumin mRNA (300 cpm; 40 cpm/pg) ($\bullet\text{---}\bullet$) or ^{14}C-unique-sequence DNA ($\circ\text{---}\circ$) were mixed with 110 ug of sheared chicken DNA (300-400 nucleotide length) in a total volume of 10 μl and incubated at 68°. Aliquots were taken at various times and the extent of hybridization was determined with S-1 nuclease. The molar ratio of ovalbumin genes to ovalbumin mRNA in this experiment is 10:1.

number of mRNA sequences by methods that are independent of the ability of the mRNA to be detected in a lysate translation assay. Such a method is useful to study instances in which mRNA may not be translatable, as well as cases in which the number of mRNA molecules is below the level of detection by the lysate system.

Figure 10 shows an analysis of the number of ovalbumin mRNA sequences in total RNA from withdrawn chicks and other tissues. In such experiments a constant amount

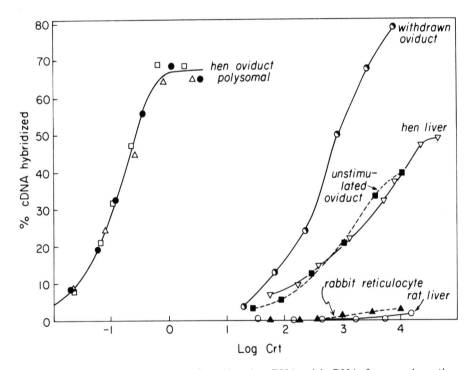

Figure 10. Hybridization of ovalbumin cDNA with RNA from various tissues. Withdrawn oviduct RNA (o---o) 100.2 A $_{260}$/ml; unstimulated oviduct RNA (■---■) 129.8 A $_{260}$/ml; rabbit reticulocyte RNA (▲---▲) 102 A $_{260}$/ml; rat liver RNA (o---o), and hen liver RNA (△---△) 295 A $_{260}$/ml were hybridized to 20 pg of cDNA. Three separate experiments are plotted for hen polysomal RNA (□---□, ●---●, △---△). The RNA preparations were washed extensively with 3M sodium acetate to remove DNA. The hen liver RNA was also treated with DNase. All hybridizations were done in a total volume of 10 μl, except for hen liver RNA which were done in 20 μl. Unincubated controls containing the RNA and all reaction components were treated with S_1 nuclease, and the percent of cDNA resistance in the absence of added RNA (2-4%) was subtracted from all values. Details are given in McKnight *et al.* (submitted for publication).

of cDNA is hybridized with RNA for increasing times or amounts of RNA (McKnight *et al.*, unpublished results). The rate of hybridization is a function of the number of ovalbumin mRNA sequences in the added RNA (Bernstiel *et al.*, 1972). (See Fig. 5 for comparison of hybridization of pure mRNA and RNA from laying hens.) In withdrawn chicks, there are a limited number of ovalbumin sequences. Such sequences have been detected even when the animals have been withdrawn from estrogen from 3 to 7 weeks. We estimate that there are approximately 78,000 mRNA sequences per cell in the laying hen and approximately 20-60 per tubular gland cell in the withdrawn chick. The estimate

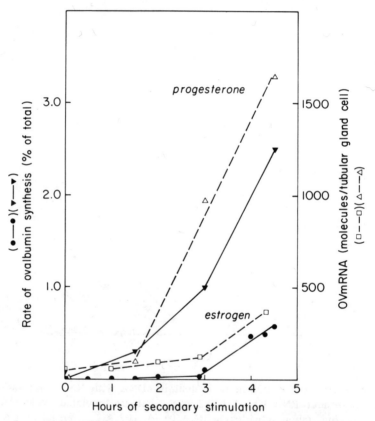

Figure 11. Comparison of the rate of ovalbumin synthesis and number of ovalbumin mRNA sequences after secondary stimulation with either estradiol benzoate or progesterone. The rate of ovalbumin synthesis (expressed as percent of total protein synthesis) was measured by standard immunologic techniques (Palmiter *et al.*, 1973). 3 mg. of each hormone was administered in sesame oil by intramuscular injection in the upper leg. The amount of ovalbumin mRNA sequences is calculated from the Crt ½ values for cDNA hybridization to RNA preparations obtained at the time points indicated. See McKnight *et al.* (submitted for publication) for details.

of 78,000 mRNA sequences per tubular gland cell agrees with the estimate of Palmiter, based on calculations of rates of ovalbumin synthesis and assumptions in transit times for ovalbumin mRNA (Palmiter, 1973). Figure 11 shows results of cDNA-RNA hybridizations to quantitate the number of ovalbumin mRNA sequences during the first hours after secondary hormone stimulation in the withdrawn chick by both estrogen and progesterone. The data has been converted to ovalbumin mRNA sequences per tubular gland cell and compared with ovalbumin synthesis as detected in intact fragments from the same tissue used for isolation of the DNA. A major increment in mRNA sequences is not detected until 3-4 hours after secondary stimulation with estrogen, whereas, the increment occurs after 60-90 minutes after progesterone administration. We believe that the lag following estrogen administration is a property of the response of the tissue, since using various routes of administration (subcutaneous, as well as intravenous and intraperitoneal) and forms of estrogens (estradiol benzoate, estradiol, diethylstilbesterol) at varying doses does not alter the 3-hour lag period. In the oviduct of the withdrawn chick we are unable to detect ovalbumin synthesis, although ovalbumin mRNA sequences are present. The increment in ovalbumin mRNA sequences corresponds well with the first ability to detect ovalbumin synthesis in oviduct fragments, irrespective of the difference in the lag period following administration of progesterone or estrogen. Palmiter (personal communication) has recently found that estrogen saturates chromatin sites within 30 minutes of secondary administration, although ovalbumin synthesis occurs only after 3 hours. Our results with cDNA indicate that this lag period is not related to post-transcriptional processing of untranslatable ovalbumin mRNA, and hence must involve events that allow for more rapid transcription of the DNA. We, therefore, conclude from these results that the amount of ovalbumin synthesized is directly related to the number of ovalbumin mRNAs. A similar conclusion has been made previously using an ovalbumin mRNA assay based on translation in a rabbit reticulocyte lysate system (Rhoads *et al.,* 1973).

Figure 10 also shows hybridization results with RNA preparations from other sources. Rat liver and rabbit reticulocyte RNA show no hybridization at time-concentrations 100 times those used for the detection of ovalbumin mRNA sequences in the withdrawn chick, indicating that there is no non-specific reaction of RNA with cDNA. However, it is interesting to find that the RNA from both chick liver and oviduct from unstimulated oviduct (virginal chicks) appears to contain a limited number of RNA sequences that hybridize with the cDNA. By a variety of criteria previously stated we believe that the cDNA is highly specific for ovalbumin mRNA sequences. Consequently, we conclude that there are, in fact, a few ovalbumin mRNA sequences (approximately 1-2 per cell) in tissues that are not specialized for ovalbumin synthesis. This finding is consistent with the concept that all genes are partially "leaky" as has been found in prokaryotes (Morse and Yanofsky, 1963).

Search for a Precursor to Ovalbumin mRNA

Although administration of estrogen to chicks results in the accumulation of

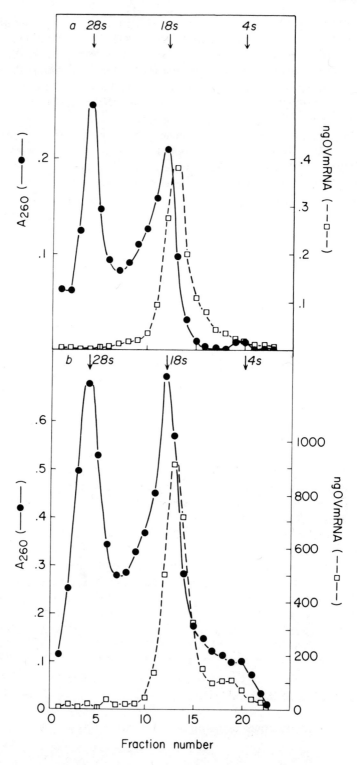

ovalbumin mRNA, there is no direct evidence that the hormone increases the rate of mRNA synthesis, although this is clearly a likely possibility. One site for action of steroid hormones might be related to an effect on the packaging and/or transport of ovalbumin mRNA precursor in the nucleus. Darnell *et al.* (1973) have proposed that mRNAs are synthesized as large molecular weight precursors, so-called heterogeneous nuclear RNA (HnRNA), with subsequent addition of poly(A) at the 3' OH end, cleavage of the product to the size of mRNA, and transport into the cytoplasm. Firtel and Lodish (1973), in contrast, have found that in *Dictostelium discoidium* the putative mRNA precursors are nearly identical in size to the cytoplasmic mRNAs. Various workers have demonstrated the existence of globin mRNA sequences in a high molecular weight fraction of duck erythrocytes, either by translation assay (Ruiz-Carrillo *et al.*, 1972) or by sequence homology with globin cDNA (Imaizumi *et al.*, 1973). However, in view of the extreme propensity for mRNA to aggregate (Strauss *et al.*, 1968), the results indicating high molecular weight mRNAs must be questioned.

We have examined high molecular weight RNA for the presence of ovalbumin mRNA sequences and have found no evidence for their existence. We can, however, readily generate artifactual aggregation of ovalbumin mRNA in RNA extracted by standard phenol-chloroform techniques when care is not taken to insure complete disaggregation. Figure 12 shows one such experiment indicating a lack of high molecular weight ovalbumin RNA sequences that can hybridize to cDNA. We have chosen to study this problem with withdrawn chicks at a time (4.5 hours) after secondary stimulation with estrogen, when there is active accumulation of ovalbumin mRNA, but also at a time when the content of ovalbumin mRNA is considerably lower than in the laying hen, and hence at a time when a small amount of precursor is unlikely to be obscured by the normal distribution of ovalbumin mRNA on a sucrose gradient, as would be the case in the laying hen.

Figure 12 shows the distribution of ovalbumin mRNA sequences detected by hybridization with cDNA. The RNA has been extracted from total tissue by phenol-chloroform extraction, with subsequent heating at dilute concentrations for 10 minutes

←————————

Figure 12. Size of ovalbumin mRNA sequences during early secondary estrogen stimulation. (a) 1.2 A_{260} units of total RNA from chicks given a secondary stimulation with estrogen for 4.5 hr was heated at 65° for 15 min to accomplish complete disaggregation of mRNA and sedimented on a 5-20% sucrose gradient for 6 hr. Fractions (0.5 ml) were collected. The A_{260} measured (●---●), and after ethanol precipitation, the total RNA in each fraction was hybridized to cDNA for 24 hr (□---□). (b) 3.2 A_{260} units of hen polysomal RNA sedimented as described above. Two percent of the DNA in each fraction was hybridized to cDNA for 20 min.

Ovalbumin mRNA sequences are quantitated as ng ovalbumin mRNA, using similar hybridization conditions and pure ovalbumin mRNA as standard. Details are presented by McKnight and Schimke (1974).

at 65° before application to a sucrose gradient. Under such circumstances, we have found that ovalbumin mRNA aggregates are effectively disrupted. In the withdrawn chick, 4.5 hours after secondary estrogen stimulation, all of the ovalbumin mRNA sequences are of the same size as ovalbumin mRNA sequences that are present in the hen, equalling the size of the polysomal mRNA. We estimate that if the half-life of the nuclear mRNA precursor were 10 minutes or greater, as suggested for globin mRNA by Imaizumi *et al.* (1973), approximately 16% of the ovalbumin mRNA sequences should be nuclear, and thus readily detected by our assays *if* they were of a size greatly exceeding that of the cytoplasmic ovalbumin mRNA. In this experiment, we would be able, in fact, to detect ovalbumin mRNA sequences in a large-size precursor if their half-life was of the order of 1 to 2 seconds. Therefore, we conclude that there is no evidence for the existence of a high-molecular weight ovalbumin mRNA precursor, and hence it is difficult to entertain a model for steroid hormone action that is involved in the packaging of a specific mRNA precursor.

Discussion

Our analysis of the mechanism(s) of regulation of specific protein synthesis as affected by steroid hormones has utilized a variety of techniques involving specific immunoprecipitation of proteins and polysomes, and DNA-RNA, and DNA-DNA hybridizations. The limiting technical problem is obtaining highly pure specific mRNA, since this material is necessary for any subsequent analysis by hybridization techniques. The immunoprecipitation techniques we have developed provide the most likely general means for such isolation, since the use of methods dependent on the size of a mRNA, *i.e.* sucrose gradients or preparative acrylamide gels, will result in isolation of all mRNAs of a specific size class. In addition to the binding of ovalbumin antibody with nascent ovalbumin chains as discussed in this paper, we have been successful with this general technique for the isolation of rat serum albumin polysomes (Shapiro *et al,* 1974), and oviduct conalbumin (Payvar and Schimke, unpublished results). A similar technique has been employed with immunoglobin light chains (Schaecter, 1974). Thus, if an antibody can be obtained that reacts specifically with nascent chains, there is no theoretical reason that the specific polysomes, and hence specific mRNA, cannot be isolated. At the present time, we do not know how generally applicable such methods will be for proteins that constitute less than 1% of proteins synthesized by a tissue (albumin and conalbumin both constitute approximately 10% of protein synthesized), and likely the technical challenges will be greater.

Our major conclusions, to date, are that a primary effect of estrogen is to regulate the content of ovalbumin mRNA, whether assayed by the translation assay or by nucleic acid hybridization. We do not find evidence for activation of a masked form of the mRNA, as is found for instance, with the increase in protein synthesis following fertilization of sea urchins (Skoultchi and Gross, 1973). How do estrogens regulate ovalbumin mRNA content? Firstly, we can conclude that gene amplification is not involved (Fig. 8 and 9). We must, therefore consider effects on some aspect of either the rate of synthesis or rate of degradation of the mRNA. We have no direct evidence on this matter, since only in the intact animal can we obtain estrogen induction of ovalbumin synthesis. Because of this, we are unable to undertake definitive studies on whether the hormone affects only the rate of synthesis of ovalbumin mRNA, only the rate of its degradation, or both. Our available evidence does not support a hormone effect on a step involving the conversion of a large molecular weight ovalbumin mRNA precursor to its cytoplasmic form, inasmuch as we are unable to detect such a precursor during early secondary stimulation with estrogen (Fig. 12).

Although it is likely that a major effect is on the regulation of mRNA synthesis, it is equally important to consider effects of steroid hormones on the rate of mRNA degradation. Recently Palmiter and Carey (Skoultchi and Gross, 1973) have shown that after rapid withdrawal of estrogen from chicks stimulated with a pellet form of estrogen that can be completely removed and hence hormone levels decrease rapidly, ovalbumin mRNA content as measured with a translation assay decreases with a half-life of 3 hours, as compared to a half-life of some 30 hours (Palmiter, 1973) in the hormone-treated bird. Thus, one can conclude that estrogens do, in fact, partially regulate ovalbumin mRNA levels by regulating the rate of degradation.

In summary, then, in the oviduct system, we have begun to dissect the regulation of specific protein synthesis at the molecular level, and have answered a limited number of questions. However, the system is extremely complex and a full understanding of the mechanisms whereby the different steroid hormones interact to regulate oviduct differentiation and function is clearly in its infancy. The use of hybridization in probes will be most useful in the development of this understanding.

REFERENCES

Baltimore, D., Jacobson, M.G., Asso, J. and Huang, A.S. (1969). Cold Spring Harbor Symposium Biol. *34*, 741.

Berns, T.J.M., Schrewis, A.M., Van Kraaikamp, M.W.G. and Bloemendal, H. (1973). *Eur. J. Biochem. 33*, 551.

Birnstiel, M.L., Sells, B.H. and Purdom, I.F. (1972). *J. Mol. Biol. 63*, 21.

Bishop, J.M. (1972). *Karoliuska Symp. on Research Methods in Repro. Endo.* 5th Symposium, p. 247.

Brawerman, G., Mendecki, J. and Lee, S.Y. (1972). *Biochemistry 11*, 637.

Breindl, M. and Galliwitz, D. (1973). *Eur. J. Biochem. 32*, 381.

Commerfold, S.L. (1971). *Biochem. 10*, 1993.

Cox, R.F. and Sauermein, H. (1970). *J. Exptl. Cell Res. 61,* 79.

Darnell, J., Jelinek, W.R. and Malloy, G.R. (1973). *Science 181,* 1215.

Firtel, R.A. and Lodish, H.F. (1973). *J. Mol. Biol. 79,* 295.

Faras, A.J., Taylor, J.T., McDonnell, J.P., Levinson, W.E. and Bishop, J.M. (1972). *Biochem. 11,* 2334.

Imaizumi, T., Diggleman, H. and Scherrer, K. (1973). *Proc. Nat. Acad. Sci. USA 70,* 1122.

Mach, B., Faust, C. and Vasalli, P. (1973). *Proc. Nat. Acad. Sci. USA 70,* 451.

Marcus, A., Efron, D. and Weeks, D.P. (1973). *Methods in Enzymology,* Vol. 30, eds. Maldave, K. and Grossman, L. (New York: Academic Press), p. 749.

McKnight, G.S., Pennequin, P. and Schimke, R.T. (Submitted for publication.)

McKnight, G.S. and Schimke, R.T. (1974). *Proc. Nat. Acad. Sci.* (In press.)

Morse, D.E. and Yanofsky, C. (1968). *J. Mol. Biol. 38,* 447.

Oka, T. and Schimke, R.T. (1969a). *J. Cell. Biol. 41,* 816.

Oka, T. and Schimke, R.T. (1969b). *J. Cell. Biol. 43,* 123.

O'Malley, B.W., McGuire, W.L., Kohler, P.O. and Korenman, S.G. (1969). *Rec. Prog. Horm. Res. 25,* 105.

Palacios, R., Palmiter, R.D. and Schimke, R.T. (1972). *J. Biol. Chem. 247,* 2316.

Palacios, R., Sullivan, D., Summers, N.M., Kiely, M.L. and Schimke, R.T. (1973). *J. Biol. Chem. 248,* 540.

Palmiter, R.D. (1973). *J. Biol. Chem. 248,* 8260.

Palmiter, R.D. and Carey, N.H. (1974). *Proc. Nat. Acad. Sci. 71,* 2357.

Palmiter, R.D., Christensen, A.K. and Schimke, R.T. (1970). *J. Biol. Chem. 245,* 833.

Palmiter, R.D. and Gutman, G.A. (1972). *J. Biol. Chem. 247,* 6459.

Palmiter, R.D., Palacios, R. and Schimke, R.T. (1972). *J. Biol. Chem. 247,* 3296.

Palmiter, R.D. and Schimke, R.T. (1973). *J. Biol. Chem. 248,* 1502.

Palmiter, R.D. and Wrenn, J. (1971). *J. Cell. Biol. 50,* 598.

Rhoads, R.E., McKnight, G.S. and Schimke, R.T. (1971). *J. Biol. Chem. 246,* 7407.

Rhoads, R.E., McKnight, G.S. and Schimke, R.T. (1973). *J. Biol. Chem. 248,* 2031.

Ruiz-Carillo, A., Beato, M., Schultz, G., Feigelson, P. (1972). *Biochem. Biophys. Res. Commun. 49,* 680.

Schaecter, I. (1974). *Biochemistry 13,* 1875.

Schimke, R.T., Palacios, R., Sullivan, D., Kiely, M.L., Gonzalez, C. and Taylor, J.M. (1973). *Methods in Enzymology,* Vol. 30, eds. Maldave, K. and Grossman, L. (New York: Academic Press), p. 631.

Shapiro, D.J. and Schimke, R.T. (1975). *J. Biol. Chem. 250,* 1759.

Shapiro, D.J., Taylor, J.M., McKnight, G.S., Palacios, R., Gonzalez, C., Kiely, M.L., and Schimke, R.T. (1974). *J. Biol. Chem.* (In press.)

Skoultchi, A. and Gross, P.R. (1973). *Proc. Nat. Acad. Sci. 70,* 2840.

Strauss, J.H.Jr., Kelley, R.B. and Sensheimer, R.L. (1968). *Biopolymers 6,* 793.

Sullivan, D., Palacios, R., Stavnezer, J., Taylor, J., Faras, A., Kiely, M., Summers, M., Bishop, J. and Schimke, R.T. (1973). *J. Biol. Chem. 248,* 7530.

THE INSECT CHORION: PROGRAMMED EXPRESSION OF SPECIFIC GENES DURING DIFFERENTIATION

Fotis C. Kafatos

The Biological Laboratories
Harvard University
Cambridge, Massachusetts 02138

Introduction

The essence of differentiation is differential gene expression. Different cell types in a multicellular organism are different in that they express different sets of genes—muscle cells the battery of genes coding for muscle proteins, pancreatic cells the battery of genes coding for digestive enzymes, etc. Gene expression is usually monitored by the rate of synthesis of the corresponding protein, although other indices may also be used (*e.g.,* rate of specific mRNA synthesis, accumulation of specific mRNA or protein, etc.). The differential gene expression which marks development has the following distinctive features.

First, the expression of developmentally regulated genes is not simply turned on and off in a step function, but instead varies quantitatively over time in a predictable manner. Thus, we must explain not only the selection of specific genes, but also the developmental kinetics of their expression. Figure 1 shows typical kinetics of specific protein synthesis for highly differentiated cells. Several phases of differentiation can be distinguished, according to the changing rate of specific synthesis (Rutter *et al.,* 1968; Kafatos, 1972). Even the "fully differentiated" phase (Phase II; Fig. 1) is characterized by a rate of specific synthesis which is changing, in many systems increasing continuously for several days.

Second, the developmental kinetics of expression are asynchronous for various genes of the differentiation-specific set. As an example we may consider the puffing response of *Drosophila* salivary glands exposed to ecdysone (Fig. 2). In this case, gene expression is manifested more directly, presumably at the level of RNA synthesis. Certain chromosomal loci respond to the hormone by decreasing their puffing, others begin puffing within minutes of treatment, still others show puffing which is delayed by hours.

In addition to the lag, the kinetics of puff growth and decline, as well as the maximal size of the puff, vary for different loci (Ashburner *et al.*, 1974). Thus, we are faced with a complex program of developmental kinetics of gene expression.

Third, programs of differential gene expression are typically autonomous and stable. Although they may be initially triggered externally (*e.g.*, by a hormone), once set in motion they become independent. The puffing response is again a good example (Ashburner *et al.*, 1974). A few early puffs are directly responsive to the ecdysone titer, but the rest (and even the turning off of early puffs) merely depend on the earlier effects of ecdysone.

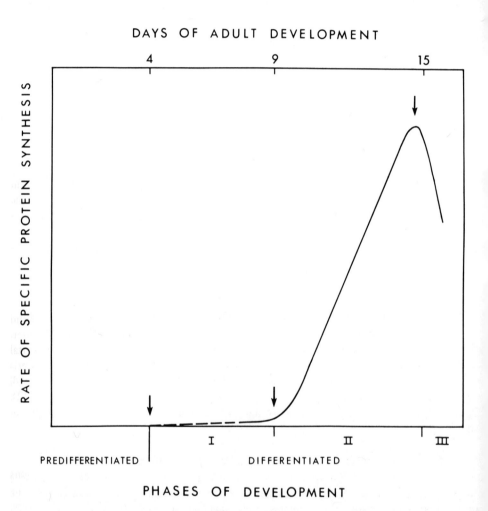

Figure 1. Developmental kinetics of gene expression, as monitored by the changing rate of synthesis of the differentiation-specific protein. The upper abscissa gives the time scale in the silkmoth galea, which produces cocoonase zymogen (from Kafatos, 1972).

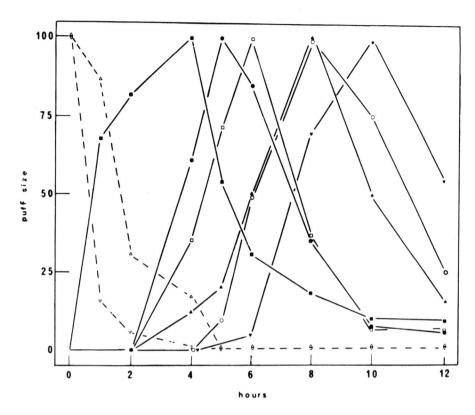

Figure 2. Asynchronous developmental kinetics of gene expression. The puffing program of *Drosophila* salivary gland chromosomes is shown, following ecdysone treatment. Each curve represents the puffing activity of a separate locus (from Ashburner *et al.*, 1974).

In order to understand cell differentiation at the molecular level, it will be important to understand the regulatory mechanisms which support stable, internally coordinated, temporal and quantitative programs of differential gene expression. In my laboratory we are working on this problem using as a model system the formation of the proteinaceous eggshell of insects. I will review briefly the information accumulated by my colleagues (named in the Acknowledgments) and myself within the last four years, and will indicate our present and future directions.

The Silkmoth Eggshell

The ovaries of a developing silkmoth consist of a total of eight assembly lines, which produce mature eggs each covered with a hard proteinaceous shell. Each assembly

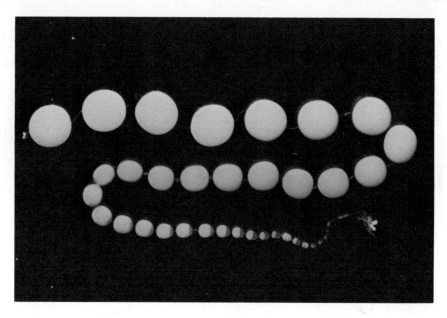

Figure 3. A silkmoth ovariole. The anterior (youngest) end is at the lower right, and the posterior (oldest) end is at the upper left. Note the cellular cord connecting the follicles in a linear array. The anterior half of each young follicle in the bottom row is occupied by the nurse cells. The follicles in the upper row are choriogenic (from Paul *et al.,* 1974).

line (ovariold) is a linear array of follicles at various stages of development (Fig. 3). Each follicle is formed at the anterior end of the ovariole and progresses backward as new follicles are added in front. Thus, the spatial dimension of the ovariole corresponds to the temporal dimension of follicle maturation. In *Antheraea polyphemus,* the species with which we have done most of our work, adjacent follicles differ by approximately 4 hours of developmental time (Paul and Kafatos, 1975).

Young follicles consist of three cell types: a single oocyte, seven nurse cells which are mitotic siblings of the oocyte, and approximately 5,000 follicular epithelial (or, more simply, follicular) cells. The giant nurse cells remain attached to the oocyte via cytoplasmic bridges, and contribute to it ribosomes, mitochondria, centrioles and other constituents (Telfer, 1975). The endopolyploid follicular cells form a single layer around the oocyte and facilitate its uptake of yolk from the blood, during vitellogenesis, by opening up intercellular channels and secreting into them yolk-binding macromolecules (Telfer and Anderson, 1968). Toward the end of vitellogenesis, the nurse cells degenerate. Yolk uptake ends, the channels between follicular cells close up, and the oocyte swells by taking up water in what is known as the terminal growth phase (Telfer and Anderson, 1968). Immediately afterwards, the follicular cells begin synthesis and deposition of the eggshell (chorion), between themselves and the oocyte. Choriogenesis lasts approximately 50 hours, and then the follicular epithelial layer breaks open (ovulation), releasing an independent, shell-covered egg which will subsequently be conveyed towards the sperm through the oviduct.

Figure 4 shows diagrams of follicles during vitellogenesis and choriogenesis. At the latter stage, since the nurse cells have degenerated, pure populations of follicular epithelial cells can be obtained by the simple expedient of cutting the follicle in half with iridectomy scissors and washing away the yolky oocyte. This is an easy operation, since the follicles are large (approximately 3 mm for the longest diameter). The dry weight of the epithelium in each follicle is approximately 0.5 mg (Paul *et al.,* 1972) and, since each ovariole usually has approximately a dozen choriogenic follicles, nearly 50 mg of choriogenic tissue (dry weight) can be obtained from each female. The dry weight of a mature eggshell is approximately 0.5 mg and 96% of it is protein. Since approximately 200 eggs are produced by each moth, 0.1 g chorion protein can be recovered from each female. The mature chorion is insoluble unless treated with a combination of a denaturant and a reducing agent. Thus, pure chorion can be obtained conveniently by cutting the eggs in half or grinding them with a mortar and pestle, and washing the yolk away with salt solutions and detergents in the absence of a reducing agent. Immature eggshells can be easily stripped with forceps from the epithelium, either in 7% propanol or preferably (in order to minimize protein loss), in 70 to 95% ethanol or in 1% H_2O_2 (pH 5).

Autoradiographic studies (Blau, 1975) suggest that chorion proteins are synthesized in the endoplasmic reticulum throughout the cytoplasm, are conveyed within minutes to widely distributed Golgi complexes, reside there for 10-15 minutes, and are then rapidly moved to the apical surface and released into the extracellular space. The half-secretion

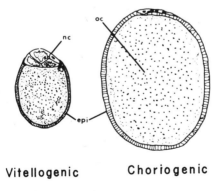

Vitellogenic   Choriogenic

Figure 4. Diagrams of a vitellogenic and a choriogenic follicle. Note the oocyte, the enveloping single layer of follicular epithelial cells, and the nurse cells which degenerate near the end of vitellogenesis. The chorion is deposited between follicular cells and oocyte.

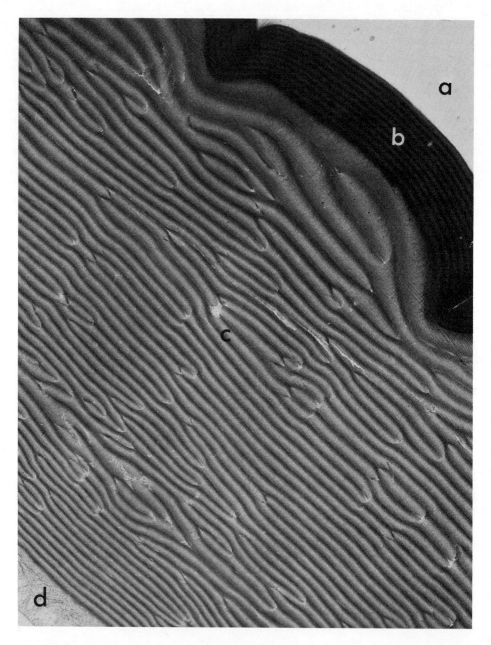

Figure 5. Transmission electron micograph of the chorion in *Bombyx mori.* a, follicular cell (underexposed because of the osmiophilia of the chorion; note the secretion granules). b, outer lamellate chorion (rich in high-cysteine proteins characteristic of this species). c, lamellate chorion (note the differences in lamellar thickness and organization, and the non-lamellar constituents). d, trabecular chorion (apparently rich in C type proteins). (From Kafatos *et al.*, 1975.)

time is approximately 20 minutes, and nearly all labeled chorion protein is extracellular one hour after a brief amino acid pulse. Autoradiography clearly indicates that during choriogenesis more than 90% of all protein synthesis in the follicular cells is devoted to production of chorion. Once extracellular, the proteins are either immobilized *in situ* (apposition), or permeate into the previously deposited chorion (intercalation), depending on the developmental stage. Newly secreted proteins are easily extracted with water. They then become water insoluble (within 3 hr), but can still be dissolved with 1% sodium dodecyl sulfate (SDS) many hours later. Final cementing together takes place gradually, through formation of disulfide bonds. Substantial resistance to SDS extraction develops only after one day from the beginning of chorion deposition, and the final one third of the chorion protein becomes crosslinked only during the 15 hours following ovulation (crosslinking monitored by the dependence of solubility on the presence of mercapto-ethanol). The extracellular morphogenesis of the chorion deserves further attention.

Protein Composition of the Silkmoth Chorion

The chorion is an ultrastructurally complex material, consisting of matrix-embedded fibers in characteristic layers, an organized system of air channels, various surface sculpturings, etc. (Fig. 5; Smith *et al.*, 1971; Kafatos *et al.*, 1975.) Paralleling the morphological complexity is the multiplicity of chorion proteins. When the proteins of a mature silkmoth chorion are dissolved with urea in the presence of a reducing agent and analyzed on an SDS-polyacrylamide gel, at least 15 components can be distinguished (Fig. 6a). The vast majority are unusually small proteins, in the molecular weight range of approximately 7,000 to 20,000. These can be distinguished into three major size classes, A, B and C (in order of increasing molecular weight). In various silkmoths, classes A and B (7,000 to 14,000 daltons) jointly account for 80 to 90% of the soluble chorion protein

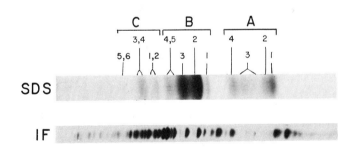

Figure 6. Chorion proteins of *Antheraea polyphemus*. Proteins fractionated by (a), top, SDS-polyacrylamide gel electrophoresis, and (b), bottom, isoelectric focusing on a polyacrylamide gel. In (a), mobility was from left to right; the major size classes and subclasses of chorion proteins are indicated. In (b), the acid (-) end was at the left for electrophoretic and focusing conditions, see Efstratiadis and Kafatos, 1975.

by mass. Minor constituents of molecular weight higher than 20,000 also exist (*e.g.*, class D, 20,000 to 30,000 daltons).

Even greater multiplicity is apparent after isoelectric focusing of the reduced, denatured proteins (Fig. 6b). More than 40 bands can be repeatably resolved on isofocusing gels, and additional minor components exist. With few exceptions, chorion proteins are unusually acidic (most showing isoelectric points in the range 4.0 to 5.5).

TABLE I

AMINO ACID COMPOSITIONS OF *A. POLYPHEMUS* CHORION PROTEINS (RESIDUES/100 RESIDUES RECOVERED)†

	Total Chorion	------------Fractionated A Proteins†† ----------			
trp	1.6	N	N	0.0	N
lys	0.4	1.1	0.8	0.9	1.1
his	0.1	0.0	0.0	0.0	0.0
arg	2.4	1.1	1.4	2.5	0.9
cys	6.1	7.7	8.0	8.4	7.7
asx	3.9	2.2	2.1	3.5	1.5
thr	2.8	2.5	2.7	2.3	2.0
ser	3.7	2.7	2.8	2.7	1.5
glx	4.7	4.3	4.2	3.3	4.7
pro	3.0	*	7.5	4.7	*
gly	32.9	33.7	32.5	35.1	40.6
ala	11.9	14.9	14.2	14.2	12.3
val	6.6	6.9	7.0	6.9	5.5
met	0.4	0.0	0.0	0.0	0.0
ile	3.8	4.1	3.8	3.6	3.9
leu	7.7	6.2	6.5	6.8	4.8
tyr	6.6	*	*	5.7	*
phe	1.5	0.6	0.5	0.6	0.6

†   From Regier, 1975, and Kafatos *et al.*, 1975.
††   Four class A protein fractions, purified twice by isoelectric focusing in a liquid column.
N   Not determined.
*   Not determined, but estimated to be 6%.

In addition to being unusually acidic and small, chorion proteins have a remarkable amino acid composition (Table I; Regier, 1975; Kafatos *et al.,* 1975). Eight non-polar amino acids make up two thirds of the residues, with glycine and alanine accounting for approximately 45% of the total. The cysteine content is high (6 to 12 molar percent in different species). Presumably, the insolubility of the chorion is based on hydrophobic interactions and disulfide crosslinks. The acidity, small size and peculiar composition of the proteins are convenient features, greatly facilitating their detection and quantitation in biosynthetic studies (see below).

The existence of multiple protein species with remarkably similar properties is noteworthy. The similarities are particularly striking within each size class (Kafatos *et al.,* 1975). For example, the carbamidomethylated A proteins of *A. polyphemus* can be purified quantitatively in a single step based on solubility; nearly all of these A proteins lack tryptophan and methionine and are particularly rich in cysteine.

The question arises whether these proteins are each the product of a distinct gene, or represent multiple post-translationally modified forms derived from a small number of precursors. Double-labeling comparisons of "old" and "newly synthesized" proteins in a single follicle indicate that the components distinguished on SDS-polyacrylamide gels are neither interconverted nor derived from a large precursor (Paul *et al.,* 1972; Kafatos *et al.,* 1975). SDS-electrophoretic analysis of nascent polypeptides shows no evidence for large precursors: the size distributions of proteins 5 minutes old and 26 minutes old on the average are superimposable, and so are the distributions of proteins 22 minutes old and 210 minutes old on the average. By contrast, experiments of similar design show that many of the components resolved by isoelectric focusing are indeed post-translationally modified (Regier, 1975). Some B proteins are blocked at the N-terminus, and this type of modification which can alter substantially the charge and not the size of a protein may be prevalent; this would explain the observation of modifications on isoelectric focusing but not on SDS gels. In any case, the modifications seem to correspond to isoelectric point alteration of components which are distinct and numerous from the outset, rather than to creation of multiple components from few precursors.

Definitive proof for the multiplicity of chorion genes comes from sequencing studies (Regier, 1975). Class A proteins of *A. polyphemus* have been purified by differential solubility, and individual components have been isolated by repeated isoelectric focusing. These proteins are very similar in terms of amino acid composition (Table I). However, sequence analysis of the N-terminal region of five of them clearly revealed that they are products of distinct, although extensively homologous genes. In addition, these studies revealed internally repeating sequences, such as the dipeptides cys-gly and gly-leu, and the tetrapeptide gly-leu-gly-tyr; it will be interesting to determine their functional significance.

The sequence homologies, together with the physical similarities, clearly suggest that chorion proteins are coded for by multigene families. According to the model formulated by Campbell *et al.* (1975), an informational multigene family includes similar but non-identical sequences of partially overlapping functions; these sequences are physically linked and evolve rapidly in species-specific clusters (coincidental evolution). Such families may be widespread in eucaryotes, since their properties permit both rapid evolution and storage of a large amount of genetic information, without a large mutational load. In addition to antibodies, mammalian keratins may also be coded for by informational multigene families (Fraser *et al.*, 1972). In connection with the informational multigene family hypothesis, further work is necessary to establish the functions of chorion proteins, their evolutionary dendrograms within and between species, and the linkage relationships of their genes. Given the wealth of developmental information about the chorion (see below), it will also be interesting to study the organization and evolution of chorion genes, with a view towards exploring the evolution of differentiation (*e.g.*, is there a tendency for recently diverged genes to be expressed coordinately?).

Programmed Changes in the Synthesis of Chorion Proteins During Differentiation

The developmental interest of the chorion system begins with the observation that various chorion proteins are produced asynchronously. Choriogenesis entails a strict program of production of many distinct proteins, each with characteristic developmental kinetics.

Figure 7 shows the proteins being synthesized by the follicular epithelium of *A. polyphemus* at various stages of differentiation (Paul and Kafatos, 1975). Follicles of known position within the ovariole (see Fig. 3) were pulse-labeled with $[^3H]$ leucine, either *in situ* (by injection into the animal) or in culture. Each follicle was then cut in half, the oocyte washed away, and the follicular epithelium plus chorion dissolved and analyzed on an SDS-polyacrylamide gel. The gel was cut into slices, and the radioactivity of each slice determined by liquid scintillation. The changing radioactivity profiles displayed in Figure 7 document the changing pattern of proteins being synthesized in the follicular epithelium, as choriogenesis proceeds. It should be recalled that the chorion proteins resolved on SDS gels are distinct components, and are not interconverted (see above). The "synthetic profiles" of Figure 7 are completely representative; identical profiles were obtained *in situ* and *in vitro*, with pulse durations of a few minutes to over an hour. These profiles are based on relative synthetic rates, which can be determined without knowledge of the specific activity of the intracellular precursor pool. Most recently, absolute synthetic rates have also been determined (Regier, 1975).

Even a preliminary examination of the synthetic profiles shows clear differences between stages, in terms of synthesis of proteins in the major size classes. Before choriogenesis (stage O), no detectable synthesis of chorion proteins (classes A to D) is occurring; the cells are synthesizing substantially larger proteins (*i.e.*, of "normal" size), which do not penetrate far into the highly cross-linked gels we use. Synthesis of such proteins rapidly declines as the follicles enter choriogenesis. First, small amounts of specific C proteins are produced (most clearly evident after labeling with glycine; Stage I_a). Synthesis of C's is prominent in early choriogenic follicles (Stages I_a to II) but becomes minor

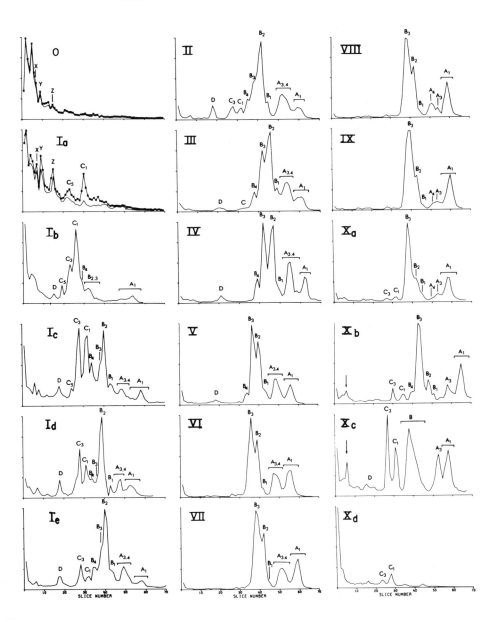

Figure 7. Differential gene expression during choriogenesis in *A. polyphemus*. Protein synthetic profiles are shown for eighteen stages of development (O to X_d), each lasting 3 hr. A developmental series of follicles was pulse-labeled with [^{3}H] leucine. Each panel represents the newly synthesized proteins in the follicular epithelium and chorion of one follicle, at the indicated stage of development. Ordinate, radioactivity. Abscissa, mobility in an SDS-polyacrylamide gel (*cf.* Fig. 6a). In panels O and I_a, *—* represents incorporation of [^{14}C] glycine. For further details, see the text. (From Paul and Kafatos, 1975.)

thereafter, except for the very end (stages X_a to X_d). The onset of substantial A and B synthesis is somewhat delayed, relative to C synthesis. B becomes the predominant class among newly synthesized proteins at stage I_d, and remains so through stage X_b. Synthesis of A's keeps pace during that time, at a rate approximately half that of B synthesis. For most of that period, production of chorion proteins (chiefly A and B's) accounts for greater than 90% of total protein synthesis in the follicular cells.

Further examination of the synthetic profiles reveals asynchronies within each class. For example, in early choriogenesis there is a progressive shift in synthesis, from subclass C_{1+2} to subclass C_{3+4}. Among B's, subclass B_2 is "earlier" than subclass B_3. Among A's, proteins in subclass A_{1+2} predominate at the very beginning (stage I_b) and at a late period (stages VII to X_b), whereas proteins in subclass A_{3+4} are dominant for much of early choriogenesis (stages I_e to V); the two subclasses are comparable at other times (stages I_c, I_d, VI, X_c). Asynchronies exist even within subclasses: note the shifting proportions of A_3 and A_4 at stages VIII to X_a.

The asynchronous kinetics of specific protein synthesis make it possible to determine the developmental state of a follicle by examining the proteins it is producing at the time. Even in mid-choriogenesis, when differences are minimal, stages can be distinguished by small but repeatable shifts in the relative prominence of labeled peaks (cf. stages V to VII). The staging system shown in Figure 7 was constructed so that the real time difference between adjacent stages would be constant; each stage corresponds to approximately 3 hours (see below). Thus, choriogenesis in any one follicle lasts approximately 50 hours. The stages, of course, are abstracted from a continuous process. At least two thirds of the stages are represented in any one female, since each ovariole usually has approximately a dozen chorionating follicles. The staging system was also related to the cumulative amount of chorion already deposited; in Figure 7, stages I, II, ..., X correspond to accumulation of 0-10%, 10%-20%, ... and 90%-100% of the final chorion dry weight, respectively.

The timing of the stages was established by a very simple procedure (Paul and Kafatos, 1975). Follicles were pulse-labeled with [3H]leucine, chased in the absence of isotope, and then pulse-labeled again, this time with [14C]leucine. The oocyte was removed, and the proteins of follicular cells plus chorion were analyzed by SDS-polyacrylamide gel electrophoresis, as usual. The [3H]protein profile revealed the synthetic stage at the beginning of the experiment, and the [14C] profile the synthetic stage at the end. Knowing the duration of the experiment and the number of stages traversed, it was possible to calculate the average duration of each stage.

The normal time course was established by in situ and in vivo experiments. In the former, [3H]leucine was injected into intact animals, and the end of the first "pulse" depended on disappearance of much of the isotope from the blood. In the latter approach, follicles were dissected, labeled briefly in culture, and then implanted into the abdomen of a host for the duration of chase. The same results were obtained with both approaches (3 hr/stage). Most importantly, in 36 successful in vitro experiments, in which the follicles were maintained in culture throughout the chase period, the average duration of the stages was again 3 hours (Fig. 8; Paul and Kafatos, 1975). This was true, even though the culture conditions led to a substantial decrease in the absolute rate of protein synthesis: the cells were able to regulate essentially normally the relative rates of synthesis of the various proteins, as they progressed along the program of choriogenesis. Thus, we conclude that the program of *relative* gene expression during choriogenesis is autonomously regulated, within the follicle, and is even insensitive to feedback from the absolute levels of protein products.

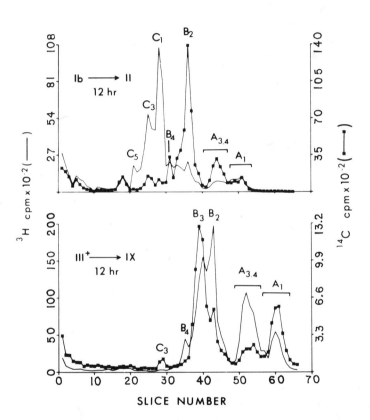

Figure 8. Changes in the protein synthetic profiles of *A. polyphemus* follicles cultured *in vitro*. Each panel represents the synthetic profiles of a single follicle, labeled at the beginning of the experiment with [^{3}H]leucine and at the end with [^{14}C]leucine. The notations at the upper left indicate the initial synthetic profile (^{3}H), the final synthetic profile (^{14}C), and the time interval between the midpoints of the two isotope pulses (hr). The average duration of each stage traversed is approximately 3 hr, the same as *in vivo* (3.0 hr for the top panel, 2.1 hr for the bottom panel. From Paul and Kafatos, 1975).

With the program of gene expression defined at the level of protein synthesis, we can begin to inquire into the nature of the regulatory mechanisms. Two main approaches seem most promising. The first is characterization of the chorion mRNAs, leading to a description of their kinetics of synthesis and accumulation, in comparison with the kinetics of protein synthesis; a basic question is whether the protein synthetic program is driven by mRNA production, or whether it is regulated post-transcriptionally. The second approach is to study mutants affecting chorion production.

Chorion mRNAs and Their Metabolism

The extreme specialization of the follicular cells, and the unusually small size of their protein products, greatly facilitated identification and purification of their mRNAs (Gelinas and Kafatos, 1973). The methods currently used in our mRNA work are discussed in detail elsewhere (Efstratiadis and Kafatos, 1975).

Basically, chorionating follicles were labeled with ^{32}P-phosphate, either *in vivo* or in culture. Cytoplasmic or polysomal RNA was extracted, and an mRNA-containing fraction was purified by a combination of sucrose gradient centrifugation (to select molecules in the general size range of monocistronic chorion mRNA) and binding to oligo(dT)-cellulose (to select poly(A)-containing molecules). This fraction was subjected to electrophoresis on a polyacrylamide gel. Autoradiography revealed that virtually the only labeled species detectable in this fraction were included in three broad zones in the 8 to 9S region. These zones were labeled zones 1, 2 and 3, in order of increasing size. Zone 2 was the most intense, and zone 3 was a minor shoulder on zone 2; our working assumption is that zones 1, 2 and 3 correspond to mRNA families for chorion proteins A, B and C, respectively.

The oligo(dT)-purified mRNA fraction was translated in the ascites and wheat germ cell-free systems to yield products with size and isoelectric point distributions similar to those of authentic chorion (Hunsley *et al.*, 1975). The correspondence was neither quantitative nor exact, as expected, given the large number of protein species involved and their post-translational modifications. Nevertheless, a substantial majority of the products were immunoprecipable with specific antibody directed towards total chorion. This suggests that contamination of the mRNA fraction with unlabeled non-chorion mRNA is low. This conclusion was reinforced by the results of reverse transcription: under conditions permitting formation of full-length reverse transcripts, the cDNAs generated fell essentially exclusively into two zones (average sizes 500 and 650 nucleotides), corresponding to mRNA zones 1 and 2+3 (Efstratiadis *et al.*, 1975).

Identification of the labeled RNA zones 1 to 3 as mRNA is consistent with their nearly exclusive occurrence on the translational machinery (97% in polysomes and ribosomes combined; Gelinas, 1974), and with their content of poly(A) (Vournakis *et al.*, 1974). Strong support for their identification as *chorion* mRNA, besides the evidence from cell-free translation, also comes from detailed size and base composition studies.

These RNAs are somewhat larger than needed to code for chorion proteins (as is the case for other mRNAs). Nevertheless, their size is greater in species with larger chorion proteins: in *Manduca sexta,* a moth whose chorion proteins are more than 5,000 daltons larger, on the average, than those of *A. polyphemus,* the size of the putative chorion mRNAs is correspondingly greater (by 50,000 to 100,000 daltons; Gelinas and Kafatos, 1973). The mRNA fraction, as well as electrophoretically purified zones 1 and 2, are relatively G+C rich, as expected from the high gly+ala content of chorion proteins (base composition of approximately 46% G+C after correction for poly(A), as compared with 35% for main band DNA; Vournakis *et al.,* 1974; Kafatos *et al.,* 1975).

Once chorion mRNAs became assayable by electrophoresis and by cell-free translation, it was possible to inquire when they are produced and accumulated, relative to the period of chorion protein synthesis (Gelinas, 1974; Kafatos *et al.,* 1975). No chorion proteins could be detected by electrophoresis and immunoprecipitation among the cell-free products, when total cellular RNA from the last pre-choriogenic follicle (stage 0) was added to the wheat germ system; chorion proteins were produced when RNA from any chorionating stage was used. Thus, no chorion mRNA translatable by the wheat germ system accumulates until *in vivo* choriogenesis actually begins. Lack of accumulation could result either from no synthesis of mRNA or from its continuous breakdown prior to the onset of choriogenesis.

Synthesis of the mRNAs was assayed electrophoretically by the following procedure. *In vivo,* ^{32}P pulse-labeled, staged follicles were homogenized. Cytoplasmic, Mg^{2+}-precipitable (Palmiter, 1974) RNA was obtained, purified by binding to oligo(dT)-cellulose, and analyzed by electrophoresis and autoradiography. The purification steps were necessary to reduce the background of non-messenger RNAs to the point where chorion mRNAs would become detectable; these steps limited the assay to cytoplasmic and polyadenylated (but not necessarily polysomal) chorion mRNAs. No chorion message production was detectable prior to the beginning of choriogenesis. By contrast, mRNA labeling was very active in the first half of choriogenesis, and continued at a declining rate until ovulation. Despite the limited resolution, there was some indication that labeling of zone 3 occurred earlier than the labeling of zones 1 and 2 (as expected from the early synthesis of C relative to A and B proteins). This time course of chorion mRNA synthesis could be contrasted with the timing of rRNA synthesis, which was extensive during late vitellogenesis, peaked at the beginning of chorion formation, and became insignificant by mid-choriogenesis.

In conclusion, the evidence indicates that chorion mRNA is produced throughout choriogenesis, and accumulates (at least in a translatable form) only during this period. It is still possible, of course, that mRNA precursors are produced prior to choriogenesis, but do not accumulate in a translatable form and do not appear in the cytoplasm as mature-size polyadenylated species. These possibilities will be tested now that full-length chorion cDNA has been produced (Efstratiadis *et al.,* 1975). Within these limits, the results suggest that chorion message production plays an important role in driving choriogenesis.

With the limited RNA resolution available, we could not evaluate the possibility that, despite contemporaneous message production, the protein synthetic program is finely tuned through translational controls. Recently, we have made substantial progress towards further resolution of the mRNAs. Chorion messages contain relatively small but polydisperse poly(A) sequences (weight average 30 to 50 nucleotides, depending on the age of the mRNA, and a range of approximately 10 to 70 nucleotides; Vournakis *et al.*, 1974). We reasoned that the range of poly(A) sizes is sufficient to cause smearing within each mRNA class, since it corresponds to 20 amino acids, which is a substantial proportion of protein size diversity within each class. Thus, we hybridized chorion mRNA with oligo(dT) and then digested the hybridized portion [poly(A)] with calf thymus RNase H (Sippel *et al.*, 1974). Electrophoresis resolved the deadenylated mRNA into at least 10 bands (Vournakis *et al.*, 1975). If these bands can now be correlated with specific protein subclasses resolved on SDS gels, the way will be open for comparing in detail the time course of mRNA and protein synthesis.

The *Drosophila* Chorion

A number of interesting chorion mutants exist in the cultivated silkworm, *Bombyx mori* (Tazima, 1964; Nadel *et al.*, 1974; Kafatos *et al.*, 1975). They appear to be pleiotropic with respect to their effect on specific chorion proteins. Each of the three mutants analyzed to date affects several of the proteins resolved on SDS and isoelectric focusing gels, mostly by altering their relative concentration in the mature chorion. One mutant in particular, Gr^{col} (grey collapsed), is associated with considerable and non-proportional deficiencies of most proteins in the eggshell; some of the proteins are almost totally absent. However, the effect appears to be exerted post-translationally, rather than at the level of regulation of the synthetic program. Even the grossly underrepresented proteins are produced at the right time and at normal relative rates, accumulate, and then disappear. The indications are that these proteins are degraded, mostly intracellularly; perhaps the mutation affects some step in secretion, resulting in intracellular accumulation and subsequent degradation of some proteins.

For studying the genetics of regulation of a protein synthetic program in a higher eucaryote, *Drosophila* is the obvious organism of choice. Indeed, a number of mutants producing fragile or abnormal chorion are already available, having originally been collected as female steriles. In a comprehensive study of recessive sex linked fs mutants, Mohler (private communication) found that 11 of 56 cistrons thus far located may affect the chorion. One of the female sterile, chorion-defective *Drosophila* strains (8-854) was studied by Margaritis (1974). The chorion mutation was mapped in the same region of the second chromosome as the female sterility mutation, suggesting that this strain is not a double mutant. Ultrastructural studies showed that two of the three eggshell zones are formed incompletely, and yet appear to require abnormally long periods of time for their formation. Whether or not this turns out to be a real regulatory mutant, it is clear that the control of chorion formation in *Drosophila* is amenable to detailed genetic analysis.

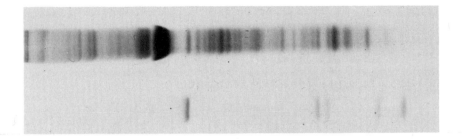

Figure 9. The proteins of *Drosophila* chorion and ovary. Chorion (endochorion plus exochorion; bottom) compared with total ovarian proteins (top). Electrophoresis from left to right, on a stacking SDS-polyacrylamide gradient slab gel (T = 7 to 14% acrylamide, C = 6% Bis). The most prominent ovarian protein is yolk. The largest prominent chorion protein has a molecular weight of 41,000 daltons, the smallest has a molecular weight of approximately 10,000. See also Petri *et al.*, 1975b.

Biochemical studies of chorion formation in *Drosophila* had to overcome two important obstacles: the complete insolubility of the mature eggshell, and the small size of the follicle. These obstacles were overcome by the development of a mass dissection procedure which, starting from a population cage of flies, yields 10^5 chorionating follicles in approximately one hour (Petri *et al.*, 1975a). From this follicle preparation, mature but not yet irreversibly cross-linked eggshells can be recovered. Figure 9 shows the major chorion proteins of *Drosophila*, analyzed by electrophoresis on an SDS-polyacrylamide gel. The proteins are quite different from silkmoth chorion proteins, both in size and in amino acid composition, but are again distinctive as compared to general follicular proteins (Kafatos *et al.*, 1975; Petri *et al.*, 1975b). The eggshell proteins are each produced with distinctive developmental kinetics (Petri *et al.*, 1975b). Thus, the essential biochemical features of the silkmoth chorion system are duplicated in the fruit-fly. We feel confident that new insights concerning the regulation of differential gene expression during development will be forthcoming from a combined genetic and biochemical study of choriogenesis in *Drosophila.*

Acknowledgments

I thank my past and present colleagues whose work I have summarized here, for making the life of the laboratory so positive and for allowing the use of unpublished results: H. Blau, A. Efstratiadis, R.E. Gelinas, M.R. Goldsmith, J.R. Hunsley, M. Koehler, L. Margaritis, G.D. Mazur, P.B. Moore, M. Nadel, M. Paul, W.H. Petri, J.C. Regier, J.N. Vournakis, and A. Wyman. Similarly, I thank my secretary, Linda Lawton, and those who contributed technical assistance: B. Baker, L. Delong, V. Raidl, J.A. Jordan and B. Klumpar. Our work has been supported by NSF (GB-35668X), NIH (5-T01-HD-94701) and the Rockefeller Foundation (RF-73019).

REFERENCES

Ashburner, M., Chihara, C., Meltzer, P. and Richards, G. (1974). *Cold Spring Harbor Symp. Quant. Biol. 38,* 655.

Blau, H. (1975). Ph.D. Dissertation, Harvard University.

Campbell, J., Elgin, S.C.R. and Hood, L.E. (1975). *Ann. Rev. Gen. 9.* (In press.)

Efstratiadis, A. and Kafatos, F.C. (1975) in *Methods in Molecular Biology,* Vol. 8, ed. Last, J. (New York: Marcel Dekker). (In press.)

Efstratiadis, A., Maniatis, T., Kafatos, F.C., Jeffrey, A. and Vournakis, J.N. (1975). *Cell. 4,* 367.

Fraser, R.D.B., MacRae, T.P. and Rogers, J.E. (1972). *Keratins* (Springfield, Illinois: Charles C. Thomas).

Gelinas, R.E. (1974). Ph.D. Dissertation, Harvard University.

Gelinas, R.E. and Kaftos, F.C. (1973). *Proc. Nat. Acad. Sci. USA 70,* 3764.

Hunsley, J., Gelinas, R.E. and Kafatos, F.C. (In preparation.)

Kafatos, F.C. (1972) in *Current Topics in Developmental Biology 7,* eds. Moscana, A.A. and Monroy, A. (New York: Academic Press), pp. 125-191.

Kafatos, F.C., Regier, J., Mazur, G.D., Nadel, M.R., Blau, H., Petri, W.H., Wyman, A.R., Gelinas, R.E., Moore, P.B., Paul, M., Efstratiadis, A., Vournakis, J.N., Goldsmith, M.R., Hunsley, J.R., Baker, B. and Nardi, J. (1975) in *Results and Problems in Cell Differentiation X,* ed. Beermann, W. (Berlin: Springer-Verlag). (In press.)

Margaritis, L. (1974). Ph.D. Dissertation, University of Athens, Athens, Greece.

Nadel, M., Kafatos, F.C. and Goldsmith, M. (1974). *J. Cell Biol. 63,* 238.

Palmiter, R.D. (1974). *Biochemistry 13,* 3606.

Paul, M., Goldsmith, M.R., Hunsley, J.R. and Kafatos, F.C. (1972). *J. Cell Biol. 55,* 653.

Paul, M. and Kafatos, F.C. (1975). *Develop. Biol. 42,* 141.

Petri, W.H., Wyman, A.R., Kafatos, F.C. (1975a) in *The Genetics and Biology of Drosophila,* Vol. II, eds. Wright, T. and Ashburner, M. (London: Academic Press). (In press.)

Petri, W.H., Wyman, A.R., Kafatos, F.C. (1975b). (In preparation.)

Regier, J. (1975). Ph.D. Dissertation, Harvard University.

Rutter, W.J., Kemp, J.D., Bradshaw, W.S., Clark, W.R., Ronzio, R.A. and Sanders, T.G. (1968). *J. Cell. Physiol.* *72*, 1.

Sippel, A.E., Stavrianopoulos, J.G., Schutz, G. and Feigelson, P. (1974). *Proc. Nat. Acad. Sci. USA* *71*, 4635.

Smith, D.S., Telfer, W.H. and Neville, A.C. (1971). *Tissue and Cell 3*, 477.

Tazima, Y. (1964). *The Genetics of the Silkworm* (London: Logos Press and Englewood Cliffs, New Jersey: Prentice-Hall, Inc.).

Telfer, W.H. (1975). *Adv. Insect Physiol.* (In press.)

Telfer, W.H. and Anderson, L.M. (1968). *Develop. Biol. 17*, 512.

Vournakis, J.N., Efstratiadis, A. and Kafatos, F.C. *Proc. Nat. Acad. Sci. USA.* (In press.)

Vournakis, J.N., Gelinas, R.E. and Kafatos, F.C. (1974). *Cell 3*, 267.

BIOCHEMICAL ASPECTS OF JUVENILE
HORMONE ACTION IN INSECTS

*Yuzuru Akamatsu, Peter E. Dunn, Ferenc J. Kézdy, Karl J.
Kramer, *John H. Law, David Reibstein and Larry L. Sanburg*

Department of Biochemistry
University of Chicago
Chicago, Illinois 60637

Introduction

Insects undergo profound morphological changes that take place long after the embryonic stage and are thus excellent models for the study of extrinsic control of genetic programs in higher organisms. In the process of metamorphosis, undifferentiated cells organized as imaginal disks develop into complex structures (legs, wings, antennae, etc.). The genetic program that dictates the ecdysone-induced differentiation and development of the imaginal discs is regulated and modulated by the juvenile hormone (JH). A high titer of juvenile hormone prevents differentiation and rapid growth of the disks. Prior to metamorphosis the hormone titer falls, and the low level then permits development to proceed. Thus, juvenile hormone exerts an *inhibitory* effect on the course of development during larval life.

In many adult insects, on the other hand, juvenile hormone plays a *stimulatory* role in the development of the reproductive system. The hormone titer is relatively high during the early larval stages (larval instars) but decreases in the late instars preceding metamorphosis. In insects that have a pupal stage, the hormone level remains low in the pupa and rises again in the adult. Figure 1 shows an idealized scheme for the developmental process in a typical holometabolous insect (one that undergoes a complete metamorphosis with a distinct pupal stage between larva and adult) and illustrates the inhibitory effect of juvenile hormone and the release from inhibition which accompanies

---

*Symposium participant.

metamorphosis. The larval instars are separated by a molting phase during which the exoskeleton is replaced. Pulses of the molting hormone, ecdysone, unleash the extensive metabolic processes that constitute molting. The presence of juvenile hormone during the larval molts maintains larval characters by inhibiting the development of pupal epidermis and other morphological structures. Before the larval-pupal molt, juvenile hormone titer plummets to a low value. Released from hormonal inhibition, the size of the imaginal disks increases rapidly and cells formerly repressed now undergo differentiation.

Juvenile hormones are secreted by the *corpora allata,* a pair of small glands lying behind the brain. Between the brain and the *corpora allata* lie the *corpora cardiaca.* The *corpora allata* are linked to the *corpora cardiaca* and to the brain by nerves that contain axons from both neurosecretory and ordinary neurons. Thus, the insect brain, the *corpora cardiaca,* and the *corpora allata* are components of the insect neuroendocrine system.

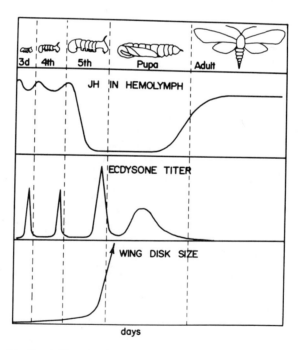

Figure 1. Idealized Hormonal Levels During Development in a Holometabolous Insect.

We now know the chemical structures of three naturally occurring juvenile hormones:

$$R_1 = R_2 = CH_3 \quad (C_{18}JH); \text{methyl 7, 11-dihomojuvenate} \qquad \text{Röller } et\ al.,\ 1967$$

R1 = R2 = CH3 (C18JH); methyl 7, 11-dihomojuvenate   Röller *et al.*, 1967

$R_1 = CH_3; R_2 = H$ (C17JH); methyl 11-homojuvenate   Meyer *et al.*, 1968

$R_1 = R_2 = H$ (C16JH); methyl juvenate   Judy *et al.*, 1973

Ref.

In addition, many synthetic analogs and even compounds that have little or no structural resemblance to the hormones have been shown to mimic hormonal activity. The impetus for the synthesis and screening of vast numbers of hormone analogs lies in the potential use of juvenile hormones for insect control. Unlike ecdysone, the juvenile hormone and many of its active analogs penetrate the insect cuticle following topical application. Applied at critical stages when endogeneous hormone titers are low, these agents lead to a disruption of normal metamorphic events and frequently result in animals intermediate between the normal stages. Such abnormal creatures usually die through failure of some vital function. Since the hormones are probably active only in a limited spectrum of invertebrates and may possess low mammalian or plant toxicity, we have the possibility of developing agents directed selectively against target organisms.

The fly in the ointment, so to speak, is the likelihood that insects already possess or will develop metabolic defenses, even toward their own hormones. Indeed, Whitmore et al. (1972) have shown that when treated with juvenile hormone, the pupa of the silkworm, *Hyalophora gloveri*, produces esterases that can hydrolyze the hormone. Analogs that are not readily hydrolyzed have been devised, but future generations of insects may learn to deal with them just as the housefly responded to DDT with a specific dehydrochlorinase (Sternberg *et al.*, 1953).

Yet the hormonal approach to insect control holds much promise and fascination. Clearly, if we want to keep ahead of the rapid pace of genetic variation that characterizes a prolific animal with a short generation time, we must be prepared to exploit new areas of vulnerability. In the case of the juvenile hormones this foresight should incorporate an examination of all aspects of juvenile hormone metabolism and function in the insect. Thus we need to consider hormone biosynthesis, storage, release, transport, catabolism, and action at target tissue, as well as other processes which we may not yet fully understand or even recognize. All of these are potential targets for insect control agents.

Along with several other laboratories, we have accepted the challenge, and this account is a progress report on what we have learned.

Biosynthesis of Juvenile Hormone

The structure of methyl juvenate is essentially that of a sesquiterpene fatty acid methyl ester with an epoxide function replacing one of the double bonds. In addition, the higher homologs contain ethyl side chains instead of the usual methyl groups. Since biosynthesis of a sesquiterpene alcohol derivative, farnesyl pyrophosphate, is well understood (Fig. 2), the special problems of juvenile hormone biosynthesis concern the understanding of the conversion of this compound to the corresponding acid, the origin of the extra carbon atoms and the methyl ester function, and the formation of the epoxide ring.

Shortly after the structure of methyl dihomojuvenate was established, Röller and his co-workers (Metzler *et al.*, 1971) began to investigate the biosynthesis of this compound in *Hyalaphora cecropia*. The adult male of this species accumulates, for unknown reasons, unusually large amounts of juvenile hormones. By injecting suspected precursors into adult males these workers established that methyl dihomojuvenate could be formed from the corresponding epoxyacid and the methyl group of methionine. Thus, it seemed likely that the methyl ester function was formed by the reaction of S-adenosylmethionine with the carboxylate group, a pathway previously observed in the bacterium *Mycobacterium phlei* (Akamatsu and Law, 1970). No evidence could be obtained for the *in vivo* formation of the hormone from other logical precursors, such as mevalonate and farnesol, although a small amount of acetate was incorporated. Furthermore, no label from the methyl group of methionine was found in the "extra" methyl of the ethyl side chain and the terminal ethyl chain. Such an origin for these carbon atoms might have been expected because of the analogy to the formation of "extra" methyl or ethyl groups in the side chains of sterols (Lederer, 1969).

A major advance in juvenile hormone biosynthetic studies was achieved by the Zoecon group, led by Siddall (Judy *et al.*, 1973). These workers were able to maintain in tissue culture the *corpus allatum-corpus cardiacum* complex* from the adult female

_____

*It is vastly easier to dissect the complex of *corpus allatum-corpus cardiacum* than to separate the glands. The *corpus cardiacum* was shown to have no capacity to synthesize juvenile hormone.

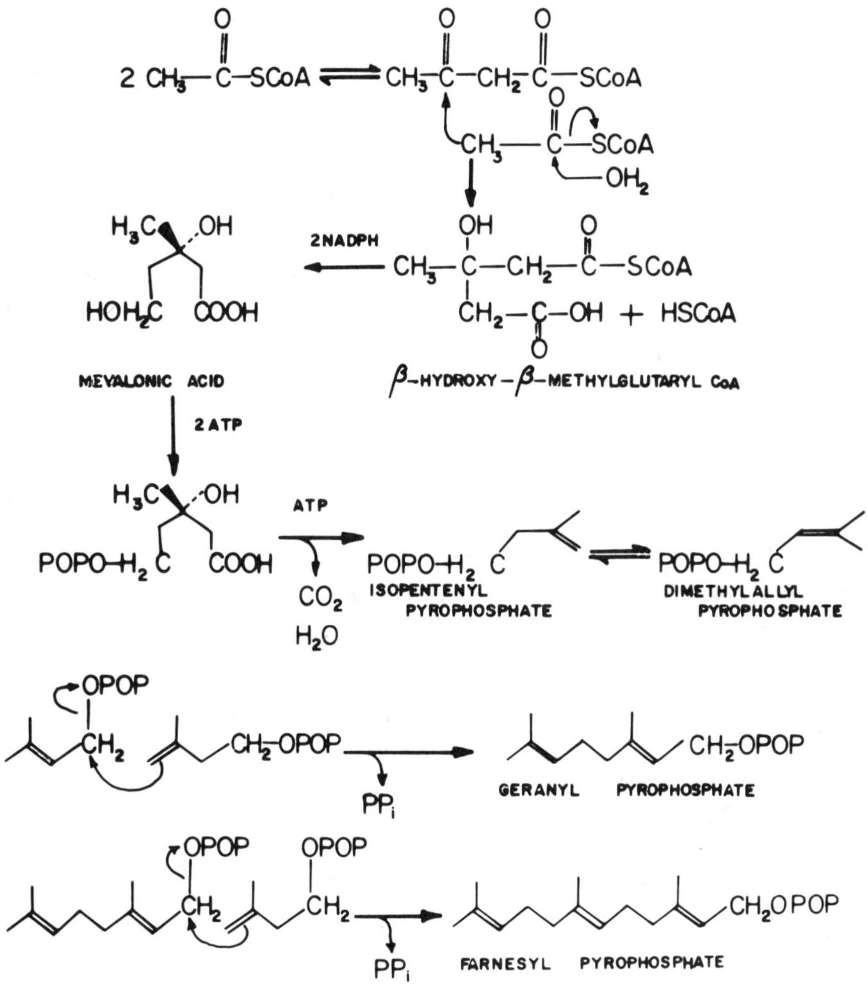

Figure 2. Biosynthesis of Farnesyl Pyrophosphate.

of the tobacco hornworm, *Manduca sexta,* and observe release of juvenile hormone into the culture medium. With this technique they confirmed the incorporation of the S-methyl group of methionine into the methyl ester function. They used this incorporation of isotopic label to identify the major juvenile hormone of *M. sexta* as methyl juvenate, a compound previously known only as an active synthetic derivative (Bowers *et al.,* 1965).

In addition, the Zoecon group (Shooley *et al.,* 1973) showed that acetate and mevalonate were incorporated and that propionate provided the ethyl side chain at the terminus of methyl homojuvenate, which is also formed in *M. sexta*. A most important conclusion drawn from these studies is that the *corpus allatum* seems to be capable of the total synthesis of the hormone from simple precursors—acetate, propionate, and methionine. A logical hypothesis is that propionate can be incorporated into a homolog of mevalonate, which then gives rise to homologous terpene precursors that are incorporated into the hormone (Fig. 3). In the known hormones the homologous olefinic alcohol pyrophosphates would have to be introduced in an orderly fashion, serving either as the first unit in the case of methyl homojuvenate, or as the first two units in the case of methyl dihomojuvenate. It is of interest to speculate that the isomeric compounds in which any or all of the three isoprene units of farnesyl pyrophosphate are replaced by homologous units may yet be found among the insect hormones.

A recent paper by Pratt and Tobe (1974) describes the organ culture of the *corpora allata* of the adult locust, *Schistocerca gregaria,* and the biosynthesis of juvenile hormone in these glands. The rates of synthesis in these cultures are quite remarkably high, at least 100-1000 times greater than in *M. sexta*. In addition to verifying methionine incorporation into the ester group, these workers have shown the conversion of farnesenic acid to methyl juvenate. The accumulation of methyl farnesenate by these glands was interpreted as evidence that biosynthesis proceeds through this compound rather than through the epoxyacid, which was only poorly incorporated. In fact, no such conclusion is warranted, for the accumulation of methyl farnesenate might be the result of a side reaction and a very low rate of conversion of this compound to the hormone. The slower incorporation of the epoxyacid may simply indicate poor uptake of this compound by the intact gland.

Because biosynthetic pathways are best investigated with cell-free systems, we decided to examine the enzymes responsible for juvenile hormone biosynthesis in homogenates of the *corpus allatum* from adult female *M. sexta*. The identification of the biosynthetic enzymes is important because any agents that could interfere with the enzymatic machinery for hormone production would cause a decrease in the hormone titers in the animal. In the larva, this would lead to precocious metamorphosis; in the adult it could prevent proper maturation of the reproductive system. Knowledge of individual enzymes should provide clues for achieving this practical goal. Furthermore, it is of interest to identify the first irreversible step in the biosynthetic process, for that is the point where we might expect to find metabolic control. In the case of cholesterol biosynthesis in mammalian liver, the first committed and rate-limiting step is the pyridine nucleotide-coupled reduction of β-hydroxy-β-methyl-glutaryl coenzyme A (HMG CoA) to mevalonic acid (Rodwell *et al.,* 1973) (See Fig. 2). Since insects do not make sterols but mevalonic acid is a hormone precursor, there may be feedback control on the HMG CoA reductase by juvenile hormone. It would be of interest to demonstrate this enzyme in the *corpus allatum,* to study its substrate specificity, and to test for its inhibition by juvenile hormones.

The minute size of the *corpus allatum* requires the use of microtechniques for

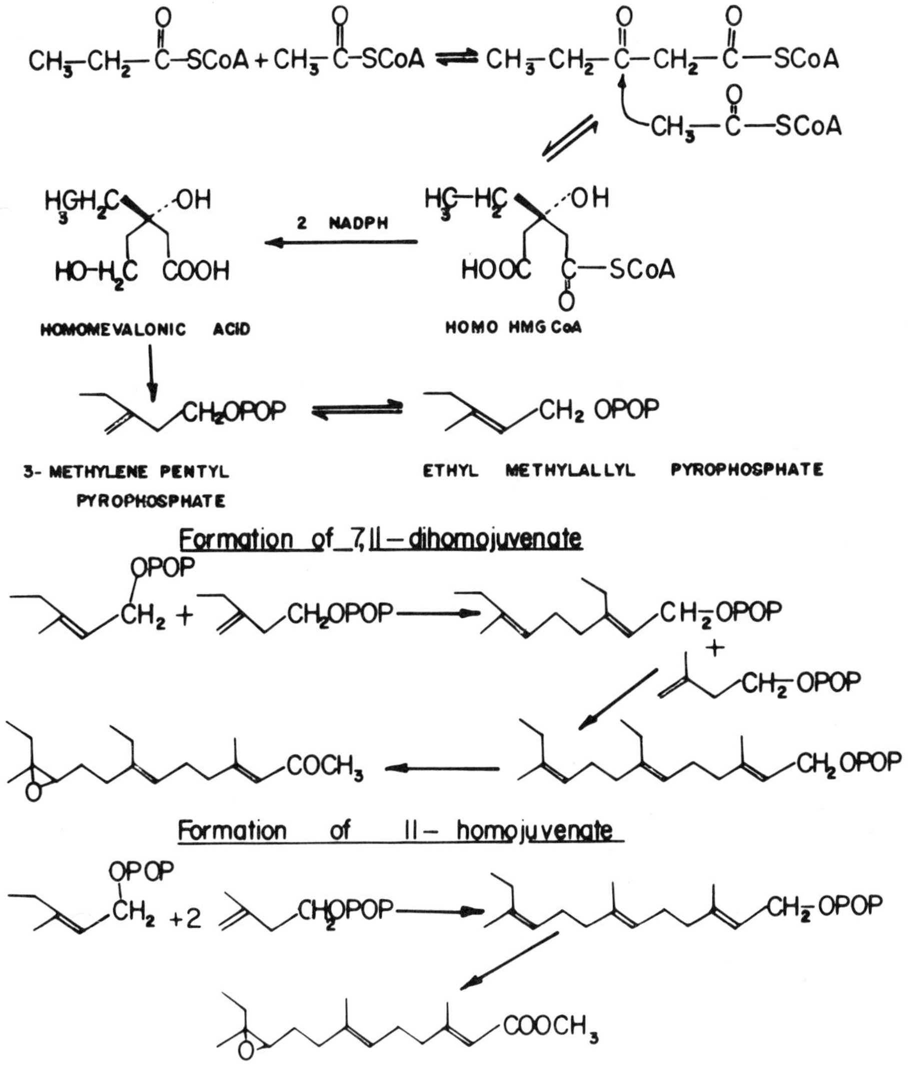

$$CH_3-CH_2-\overset{O}{\overset{\|}{C}}-SCoA + CH_3-\overset{O}{\overset{\|}{C}}-SCoA \rightleftharpoons CH_3-CH_2-\overset{O}{\overset{\|}{C}}-CH_2-\overset{O}{\overset{\|}{C}}-SCoA$$

$$CH_3-\overset{O}{\overset{\|}{C}}-SCoA$$

HGH$_3$H$_2$C ,,OH
HO-H$_2$C COOH

HOMOMEVALONIC ACID

← 2 NADPH →

H$_3$C-HC ,,OH
HOOC $\overset{}{\underset{O}{C}}$—SCoA

HOMO HMG CoA

CH$_2$OPOP ⇌ CH$_2$ OPOP

3- METHYLENE PENTYL PYROPHOSPHATE   **ETHYL METHYLALLYL PYROPHOSPHATE**

Formation of 7,11 – dihomojuvenate

OPOP
CH$_2$ + CH$_2$OPOP ⟶ CH$_2$OPOP

+

CH$_2$OPOP

COCH$_3$ ⟵ CH$_2$OPOP

Formation of 11 – homojuvenate

OPOP
CH$_2$ +2 CH$_2$OPOP ⟶ CH$_2$OPOP

COOCH$_3$

Figure 3. Suggested Pathways for Biosynthesis of Juvenile Hormones.

enzymological studies. These were provided by neurobiologists, who have devised methods for extracting enzymes from single neurons (Hall *et al.*, 1970) with a micro-homogenizer. Using this technique we are able to extract a few *corpus allatum-corpus cardiacum* complexes in less than 10 microliters of buffer and to obtain enzymatically active preparations. We have dissected the glands under aseptic conditions, prepared the homogenates in a sterile tissue culture medium containing antibiotics, and used labeled precursors of very high specific activity for long incubation times in order to achieve enough incorporation of radioactivity to characterize properly the products.

We first examined the enzymatic incorporation of the methyl group from [methyl-^{3}H]-S-adenosylmethionine into juvenile hormones (Reibstein and Law, 1973). We found that all three known hormones were produced, although production of labeled material corresponding to methyl dihomojuvenate, not previously known to occur in *M. sexta,* was highly variable and could not be reproduced consistently. In nearly every incubation methyl juvenate was the major product. When the epoxyacids were added as methyl group acceptors, hormone production was greatly stimulated (Table I). Juvenic acid (10, 11-epoxyfarnesenic acid) stimulated the production of methyl juvenate 16-fold. Tritiated dihomojuvenic acid, obtained by base hydrolysis of commercial [^{3}H] methyl ester, was also readily incorporated into methyl dihomojuvenate, and when [methyl-^{14}C]-S-adenosylmethionine was added as well, very nearly equivalent amounts of the two isotopes were incorporated into the hormone.

Unlike the relatively nonspecific bacterial enzyme (Akamatsu and Law, 1970), the methyl ester synthetase from the *corpus allatum* seems to be rather specific for sesquiterpenoid fatty acids. Oleic acid and epoxystearic acid were not converted to methyl esters by the insect enzyme, nor was 9-keto-*trans*-2-decenoic acid, a pheromone from the honeybee, *Apis mellifera.* Methyl ester synthetase was not present in *M. sexta* brain or fat body.

From this point, we have attempted to trace the pathway back toward earlier precursors. Figure 4 summarizes the possible pathways for the conversion of farnesenic acid to methyl juvenate in *M. sexta.* Because it is difficult to obtain some of the intermediates in isotopic form with high specific activity, we have tested the ability of the unlabeled compounds to stimulate incorporation of [^{3}H]-S-adenosylmethionine into the corresponding methyl esters. Addition of farnesenic acid to the homogenates resulted in very little stimulation of hormone production; instead, the methyl ester of farnesenic acid was formed (Table I, line 5). When NADPH was added simultaneously, however, a high level of labeled hormone was produced with no accumulation of methyl farnesenate (line 6). Thus, it appears that this epoxidation is catalyzed by an NADPH-coupled mixed-function oxidase, as are most known biological epoxidations. These experiments do not determine, however, whether methyl farnesenate is an intermediate in the biosynthesis of the hormone or a by-product formed only when the absence of reduced pyridine nucleotide prevents epoxidation. In order to assess the relative importance of the four reactions (nos. 2-5) shown in Figure 4, it was necessary to test labeled methyl farnesenate as a hormone precursor. Under conditions where an appreciable amount of hormone was formed from unlabeled farnesenic acid and labeled S-adenysylmethionine, we would observe no conversion of methyl farnesenate to the hormone (Table I, line 4).

Finally, we prepared ^{3}H-labeled farnesyl pyrophosphate from labeled mevalonate using hog liver enzymes (Popják, 1969). Labeled methyl juvenate was obtianed when farnesyl pyrophosphate was used as a precursor in the presence of S-adenosylmethionine and NADPH (Table I, line 8). During the same incubation, appreciable amounts of farnesenic acid and its epoxy derivative also accumulated, thereby suggesting that these compounds could be intermediates. In agreement with the results obtained with farne-senic acid as the starting material (Table I, line 6), only a small amount of methyl farnesenate accumulated under these conditions. In the absence of any added S-adenosylmethionine, only farnesenic acid and juvenic acid were produced. These findings indicate that enzymes are present in the *corpus allatum* that can cleave farnesyl

TABLE I

RADIOACTIVE LABELING EXPERIMENTS IN THE BIOSYNTHESIS
OF JUVENILE HORMONE IN HOMOGENATES OF
CORPORA ALLATA FROM *M. SEXTA* ADULT FEMALES

Precursor (pmoles)	S-adenosyl-methionine	NADPH	Products formed (pmole)			
			farnesenic acid	juvenic acid	methyl farnesenate	methyl juvenate
juvenic acid						
1 *10^2	+	-	n.d.	n.d.	n.d.	3.8
2 10^2	*+	-	n.d.	n.d.	n.d.	1.3
3 10^4	*+	-	n.d.	n.d.	n.d.	8.0
methyl farnesenate						
4 *5×10^4	-	+	n.d.	n.d.	10^4	0
farnesenic acid						
5 10^2	*+	-	n.d.	n.d.	2.9	0
6 10^2	*+	+	n.d.	n.d.	0	3.5
farnesyl pyrophosphate						
7 *10^2	-	+	1.0	0.9	0	0
8 *10^2	+	+	1.4	1.7	0.01	0.14

*Labeled compound.
S-adenosylmethionine, when present, 5.5×10^{-5}M
NADPH, when present, 10^{-3}M

pyrophosphate and convert farnesol to the corresponding acid. In summary, our results demonstrate the existence in the *corpus allatum* of enzymes capable of catalyzing four of the five conversions outlined in Figure 4.

The relative importance of the two branches of the pathway *in vivo* can be estimated by considering the yields of the different reactions, as summarized in Table I. For these arguments, one must assume that steady state was obtained for all intermediates, that the yields of the various compounds truly reflect their rates of formation and decomposition, and that side reactions are absent. With these assumptions, the approximately equal yields of farnesenic acid and the epoxyacid (Table I, lines 7 and 8) indicate that the ratios $\frac{V_1}{V_3}$ and $\frac{V_3}{V_5}$ are nearly equal. At the same time, the much lower yield of juvenile hormone indicates that V_5 is smaller than V_3 and hence V_3 is smaller than V_1. From the same experiments we can state that V_3 must be larger than $V_{2'}$ since the methyl ester does not accumulate to a significant extent and the amount of hormone produced is low. The experiment shown in Table I, line 4 demonstrates that in our experimental conditions no hormone is formed from methyl farnesenate, *i.e.* $V_4 \cong 0$. Thus, in this experiment the relative rates are in the order $V_1 \rangle V_3 \rangle V_5 \cong V_{2'}$ and $V_4 \cong 0$. The results obtained with other precursors reported in Table I are consistent with these conclusions. The yield of methyl ester reported in line 5 indicates that indeed V_2 is slower than V_3. When the addition of NADPH opens the epoxyacid branch (line 6) the ester concentration

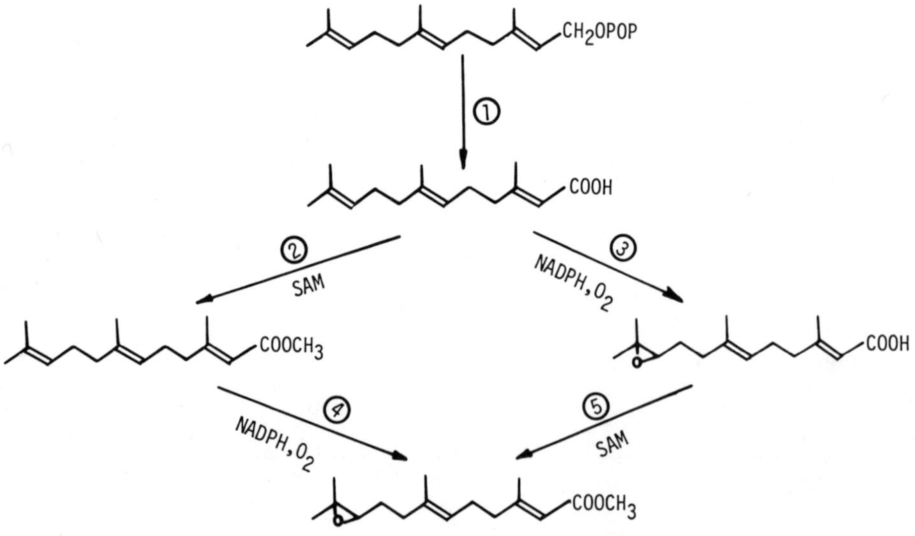

Figure 4. Biosynthesis of Juvenile Hormone From Farnesyl Pyrophosphate.

becomes negligible, because then $V_3 \rangle V_2$. If these experiments truly reflect the relative importance of the two pathways in the *corpus allatum,* then the branch going through the epoxyacid would be the predominant pathway, by at least an order of magnitude. This situation need not be the same in *Schistocerca,* but the epoxidation pathway doubtless requires molecular oxygen, and in the case of experiments with whole glands (Pratt and Tobe, 1974) oxygen may be limited since the normal access to the glands is by the trachial system, which has been severed. This might account for the accumulation of methyl farnesenate in the *Schistocerca* gland experiments, and would be consistent with all of our homogenate incubations, in which methyl farnesenate accumulates only when epoxidation is prevented by the absence of NADPH.

Together with the results of others, our experiments demonstrate that all of the enzymatic machinery for hormone synthesis is present in the *corpus allatum* and that an important enzyme, methyl ester synthetase, is absent in other insect tissues. Thus it is likely that the *corpus allatum* is the site of *de novo* hormone production, a conclusion in agreement with gland extirpation and transplantation experiments.

Release of Juvenile Hormone

It is generally believed that in holometabolous insects the *corpora allata* cease producing juvenile hormone during the pupal stage and resume synthesis in the adult (Wyatt, 1972; Gilbert and King, 1973). It is thought that secretion of the hormone is stopped in response to some signal acting through the brain by neurosecretion or neurotransmitter mediation, or both. The mechanism by which secretion would be stopped at the level of the *corpus allatum* is unknown at the present time. That is, the question remains open whether the biosynthetic enzymes disappear, or whether the actual secretion process is halted and biosynthesis is stopped because accumulated hormone exerts feedback inhibition.

Many studies have attempted to define the neuroendocrine control system in various animals in all life stages, but precious little consistent and concrete evidence has been adduced. Secretion is usually stimulated by substances produced by neurosecretory cells in the *pars intercerebralis* of the brain and secretion might be inhibited by direct innervation of the *corpus allatum* by neurons from the brain (Doane, 1973). Much of the confusion about these matters arises because most of the studies were conducted before the structures of the hormones were known and thus relied upon biological activity of the hormone to assess secretory activity. Furthermore, little is known about the specific morphological structures within the *corpus allatum* that are involved with synthesis, storage, and secretion of the hormones. The time has now arrived when more definitive experiments can be done using tissue culture, appropriately labeled hormone precursors, and rigorous analytical techniques.

Transport of Juvenile Hormone

Hormone secreted by the *corpus allatum* must pass through the hemolymph to reach the target tissues. The lipophilic nature of the juvenile hormone molecule suggested that it might be transported by lipoproteins, in a manner similar to other fatty acid esters, *e.g.* diglycerides. Chino *et al.* (1969) studied extensively the hemolymph lipoproteins in *Hyalophora* that are responsible for the transport of diglycerides from the fat body to muscle and other tissues. Whitmore and Gilbert (1972) showed that some of the lipoproteins of *H. gloveri* pupae could bind ^{14}C-juvenile hormone, and that lipoproteins from adult males contained enough hormone to be detectable by bio-assay. Binding of juvenile hormone to lipoproteins was also reported in *Tenebrio molitor* by Trautmann (1972) and in *Locusta migratoria* by Emmerich and Hartmann (1973).

These findings were rationalized on the basis of an assumed water-insolubility of juvenile hormone. This idea was popularized by Williams (1956), but in fact, the true solubility of these compounds in water had never been accurately determined. In the investigations that suggested lipoprotein carriers for JH, the hormones were administered in organic solvents. Nonetheless, Whitmore and Gilbert (1972) found that in the same experiments unbound labeled juvenile hormone was eluted from gel filtration columns in the inclusion volume. Although these authors made no note of this fact, it suggested to us that the hormone would form true, monomeric solutions. We confirmed this, and indeed, we were able to show that true aqueous solutions of juvenile hormone up to 5×10^{-5} M could be prepared and that in these solutions the hormone molecules were in monomeric form (Kramer *et al.,* 1974). Since the concentration of juvenile hormone circulating in the hemolymph of a larval insect is less than 10^{-7} M, there is no *a priori* reason why the hormone could not be in true solution.

We studied the interaction of aqueous solutions of labeled juvenile hormone with the hemolymph of *M. sexta* fifth instar larvae. Gel filtration revealed a macromolecular hormone complex, as shown in Figure 5A. Extraction of the labeled material from this complex (peak II) followed by thin-layer and gas-liquid chromatography showed that most of the hormone was still intact. In contrast, material from peak I, which contains small molecules, was shown to be degraded hormone. The presence in hemolymph of enzymes capable of hydrolyzing the methyl ester function of juvenile hormone had been observed by several investigators (Whitmore, 1972; Slade and Zibitt, 1973; White, 1972; Ajami and Riddiford, 1973; Siddall *et al.,* 1971; Riddiford and Ajami, 1973; Weirich *et al.,* 1973; Weirich and Wren, 1973). We found that prolonged incubation of the crude hemolymph with the serine esterase inhibitor, diisopropylphosphorofluoridate (DFP), resulted in complete inhibition of juvenile hormone hydrolysis. DFP-treated hemolymph incubated with labeled hormone was subjected to gel filtration, as shown in Figure 5B. Two macromolecular complexes, both containing intact hormone, were now observed. The larger of these was observed only at relatively high juvenile hormone concentrations and may be a lipoprotein. The smaller complex has a molecular weight of about 30,000 and is the more important in terms of the quantity of hormone bound. The same complexes were also observed when the hemolymph was subjected to polyacrylamide gel

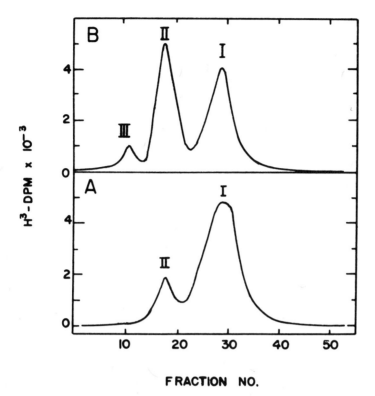

Figure 5. Gel Permeation Chromatography of *M. sexta* Hemolymph (pH 7.3, 4°, Sephadex G-100).
  A. Fifth Instar Hemolymph Incubated with 10^{-8} M JH.
  B. Fifth Instar Hemolymph Inhibited with DFP (10^{-3} M, 12 hr) Followed by Incubation with 10^{-8} M JH.

electrophoresis. The major hormone-binding protein could be detected both by the radioactivity of the bound juvenile hormone and by staining for protein by Coomassie Blue.

Estimation of the dissociation constant of the binding protein-hormone complex by rapid gel filtration gave a value of $K = 3 \times 10^{-7}$ M; the concentration of binding protein in the hemolymph was estimated by the same method to be 8×10^{-6} M. The concentration of hormone in the *M. sexta* larva would be much lower, of the order of 10^{-8} M. If the binding is a simple equilibrium then,

$$K = \frac{[H] \, [BP]}{[C]}$$

where K is the dissociation constant and [H], [BP], and [C] are the concentrations of free hormone, free binding protein, and complexed hormone, respectively. One can rearrange this equation into

$$\frac{[C]}{[H]} = \frac{[BP]}{K}$$

Under the physiological conditions specified above, the free binding protein is nearly equivalent to the total binding protein, and therefore

$$\frac{[C]}{[H]} = 25.$$

This indicates that in the fifth instar larva of *M. sexta* about 95% of the juvenile hormone would be in the form of the complex. With the same dissociation constant, the amount of hormone in the complex form could be altered by changing the concentration of the binding protein, as long as the latter remains well above the total amount of hormone.

Following a suggestion by Dr. John Katzenellenbogen, from the University of Illinois, we developed a rapid assay for binding protein by adapting methods devised for the determination of the estrogen binding protein (Katzenellenbogen *et al.*, 1973). The latter procedure uses activated charcoal to adsorb unbound ligand followed by centrifugation to remove the charcoal-ligand mixture. We found that labeled juvenile hormone is readily adsorbed to charcoal, and this fact has proved useful in a variety of problems encountered in our work with this material. Using this rapid binding assay, we proceeded to purify the binding protein.

We used gel filtration of the lyophilized and reconstituted hemolymph as the first step, for it separates various materials that interest us. The binding protein was then purified further by ion exchange chromatography or isoelectric focusing or both. By ion exchange chromatography on DEAE cellulose, two fractions with affinity for juvenile hormone were separated, and each was purified to homogeneity (Fig. 6).

Binding protein I, the major species, has been studied more extensively. It is a single polypeptide chain of molecular weight $\cong$ 30,000. Equilibrium dialysis studies with labeled hormone and pure binding protein showed only one binding site per molecule and a dissociation constant of 2×10^{-7} M, in excellent agreement with the value determined in DFP-inhibited hemolymph (Kramer *et al.*, 1974). The amino acid compositions and some of the properties of these proteins are shown in Table II.

The specificity of binding has not yet been extensively studied, but it was shown that binding of the free acid or the diol ester, which results from opening of the epoxide ring, could not be detected by the gel filtration method (Kramer *et al.*, 1974). This would mean that K is at least two orders of magnitude higher for these metabolites than for intact hormone.

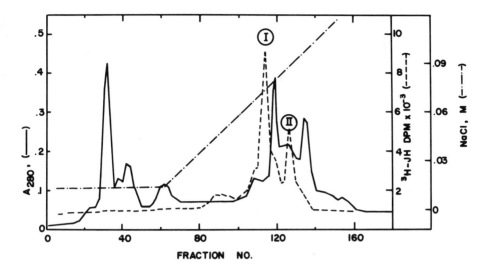

Figure 6. Biogel DEAE Ion Exchange Chromatography of Gel Permeation Fractions Containing Carrier Proteins (pH 8.3, 4°, NaCl Gradient).

We believe that one function of the binding protein is to transport the hormone from the site of synthesis to the target tissues. If, as we have shown, juvenile hormone could exist in true solution at physiological concentrations, why is a macromolecule required? We do not have a definitive answer to this question, but we have some illuminating information.

First of all, since more than 95% of the hormone in the hemolymph will be complexed to the binding protein, this complex should be important to the animal. Target cells and degradative enzymes "see" this complex, rather than free hormone. Any *in vitro* studies, as well as many *in vivo* ones, must now take the formation of the complex into account. Some hormone analogs may act by interfering with the function of the binding protein. An illustration of the importance of the binding protein as a carrier comes from our extensive studies with the hormone esterases of hemolymph, described below.

It seems possible that the carrier protein may play a protective role at target tissues, where hormone degradation is known to take place (Chihara *et al.,* 1972). Experiments by Herbert Oberlander, USDA laboratory at Gainesville, Florida, have shown that in *Plodia* imaginal disks in culture, the carrier protein strongly enhances the action of juvenile hormone (Sanburg *et al.,*1975b). Whether this is simply a protective effect, or whether the carrier protein plays some active role in transporting hormone into target cells remains to be explored.

TABLE II

AMINO ACID COMPOSITIONS OF
JUVENILE HORMONE CARRIER PROTEINS 1 AND 2

| | Amino acids per molecule | |
	CP-1[a]	CP-2[a]
Aspartic acid	31	31
Threonine	12	13
Serine	15	30
Glutamic acid	26	44
Proline	13	25
Glycine	14	31
Alanine	20	26
Cysteine[b]	10	10
Valine	19	18
Methionine[b]	3	4
Isoleucine	17	13
Leucine	20	18
Tyrosine	7	9
Phenylalanine	10	14
Lysine	17	13
Histidine	7	5
Arginine	9	12
Tryptophan[c]	2	3
MW	27965	34395

[a]Based on 31 moles of aspartic acid per mole of protein. Average of 24-, 48- and 72-hour hydrolysates. Labile amino acids extrapolated to zero time. Nearest integers of mean values presented.

[b]Determined by performic acid oxidation.

[c]Estimated from 280 nm absorbance.

Another possible function of the carrier protein remains a distinct, but untested, possibility. The hormone is a sparingly soluble lipid-like molecule which has surfactant properties, i.e. it tends to accumulate at surfaces. We have experienced difficulties in transferring very dilute solutions of the hormone for this reason. We have shown that a molecule of similar surfactant character, tripropionin, tends to accumulate at interfaces in a soluble monolayer or "surface excess" (Brockman et al., 1973). Many biologically important molecules are surfactants and undoubtedly show similar behavior. It may be detrimental to permit high local concentrations of potent biologically active compounds near cell surfaces, and carrier proteins might serve to prevent surface excesses of such materials. It is probably significant that mammals have circulating binding proteins for corticoids (Nocenti, 1968b) and for retinol (Kanai *et al.*, 1968), also sparingly soluble molecules that would be expected to have surfactant character.

Catabolism of Juvenile Hormone

Methods for deactivating juvenile hormone are important to the insect for control of hormone titers, and for dealing with the machinations of humans who attempt to turn the physiology of the insect against it. They are important to the same human pragmatists as targets for insect control, for either stimulation or inhibition of catabolism would upset the delicate hormonal balance.

The fate of labeled hormone has been followed in a variety of insects (Whitmore, 1972; Slade and Zibitt, 1973; White, 1972; Ajami and Riddiford, 1973; Siddall *et al.*, 1971; Riddiford and Ajami, 1973; Weirich *et al.*, 1973; Weirich and Wren, 1973). The predominant catabolic pathways are shown in Figure 7. Some of these processes may occur in the hemolymph, as, for example, reaction 1 in the case of the tobacco hornworm, while others occur in the fat body and in other tissues. An important point is

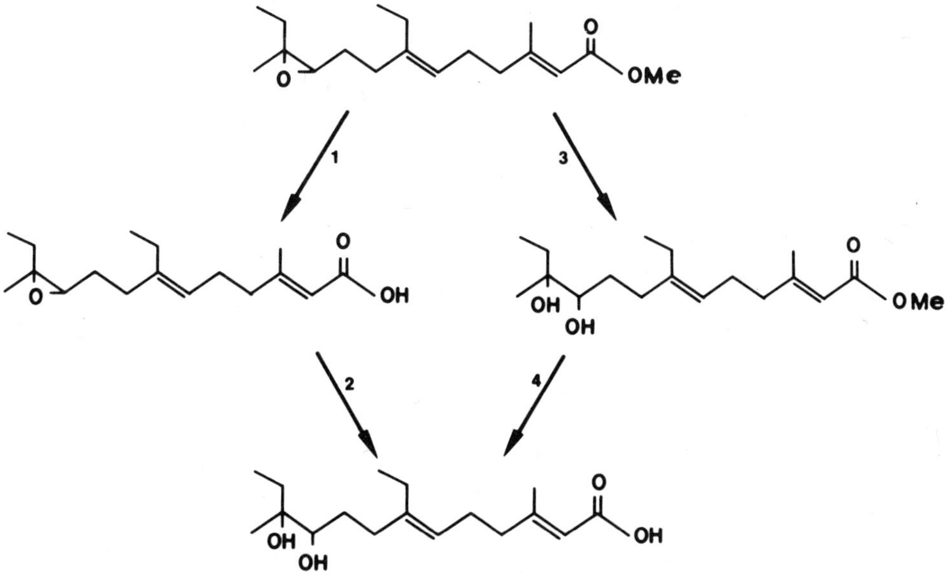

Figure 7. Initial Steps in the Degradation of Juvenile Hormone.

that all of these studies (Whitmore, 1972; Slade and Zibitt, 1973; White, 1972; Ajami and Riddiford, 1973; Siddall *et al.,* 1971; Riddiford and Ajami, 1973; Weirich *et al.,* 1973; Weirich and Wren, 1973) were done with hormone applied topically or administered in organic solvents, oils, or emulsions. How hormone administered by these modes compares with the endogenous hormone is unknown. Since in many insects the hormone could exist predominantly as a complex with carrier protein we decided to reinvestigate the catabolic reactions of juvenile hormone in hornworm hemolymph.

As the hemolymph is a complex mixture of mutually interacting systems, we first undertook the separation of its various macromolecular components. We followed the fractionation by a simple assay for hormone hydrolysis. Our studies with the carrier proteins had demonstrated the utility of activated charcoal for removing hormone from aqueous solution. We reasoned that if we used hormone labeled in the methoxy group, hydrolysis of the ester would free labeled methanol, which, by virtue of its inability to bind to charcoal, could easily be separated from the unreacted labeled hormone. Indeed, this procedure provided the basis for a rapid and reliable assay for hormone hydrolysis.

We also determined "general esterase" activity using 1-naphthyl acetate, since the hydrolysis product, 1-naphthol, can be determined directly in a recording spectrophotometer or indirectly by coupling with a diazonium salt. Using these two substrates we could judge the relative specificity of enzymes in the hemolymph for juvenile hormone. As a base line for lack of specificity, we used the rates of hydrolysis of the two substrates by hydroxide ion. It is interesting to note that the rate constant for alkaline hydrolysis of the juvenile hormone is 1000 times smaller than that for 1-naphthyl acetate. This can be rationalized on the basis that 1-naphthol is a much better leaving group than methanol and that the β-methyl group and the α,β double bond in the hormone provide shielding and resonance stabilization for the ester group. Provided that the general pathways of alkaline and enzymatic hydrolyses are the same, an increase in the specificity of an enzyme for the hormone would diminish the difference between the rates of hydrolysis of hormone and naphthyl acetate.

When fifth instar hemolymph was subjected to gel filtration chromatography, two groups of enzymes were isolated, which had the ability to hydrolyze free juvenile hormone (Fig. 8). The first of these was associated with the bulk of general esterases, while the second had only limited action on 1-naphthyl acetate. The material in this second peak of hormone hydrolase activity was fractionated further by isoelectric focusing to yield three enzymes of relatively high specificity toward juvenile hormone. The last traces of general esterase activity were removed by brief treatment with the inhibitor DFP. This material reacts with serine hydroxyls at the active site of esterases to yield nonhydrolyzable phosphate esters, thus blocking the active site from further reaction. We discovered that a few minutes of incubation with this inhibitor at low concentrations (10^{-4} M) completely inhibited the action of enzymes in Sephadex fraction I (Fig. 8) on both substrates and all general esterase activity in fraction II. On the other hand, the specific juvenile hormone hydrolase enzymes of fraction II were only slowly inhibited, even at concentrations of 10^{-3} M DFP. This fact has provided us with a specific tool for locating and measuring juvenile hormone specific hydrolases.

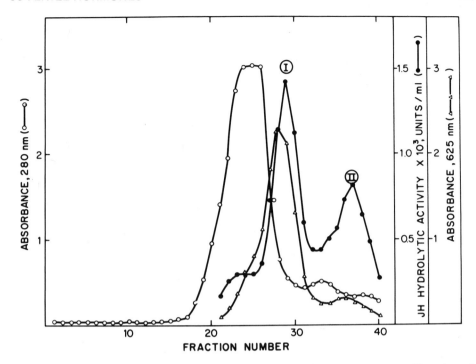

Figure 8. Gel Permeation Chromatography of Fifth Instar *M. sexta* Hemolymph (pH 7.3, 4°, 0.1 M NaCl, Sephadex G-100).   Absorbance at 625 nm is a measure of General Esterase Activity.

With the separated enzymes in hand, we now tested the action of each group on the hormone-carrier protein complex. We found that the carrier protein completely protected the hormone from hydrolysis by the enzymes of Sephadex fraction I, but not at all from those of Sephadex fraction II. Thus the specific juvenile hormone hydrolases can destroy the hormone even when it is complexed to the carrier protein.

If this is the case, how can even the complexed hormone survive in the hemolymph? We reasoned that perhaps the specific hydrolases were present only in the fifth instar, but not in the fourth or earlier instars, where high hormone levels are present. When we subjected fourth instar hemolymph to gel filtration, we indeed observed that fraction I was present, but that fraction II was nearly absent (Fig. 9). Testing of each fraction after preincubation with DFP confirmed that juvenile hormone-specific hydrolase was barely detectable.

Using the preincubation with DFP to distinguish the two esterase classes, we followed the changes in esterase level through the fourth and fifth larval instars (Fig. 10). We found that while the level of general esterase fluctuates, that of the hormone-specific

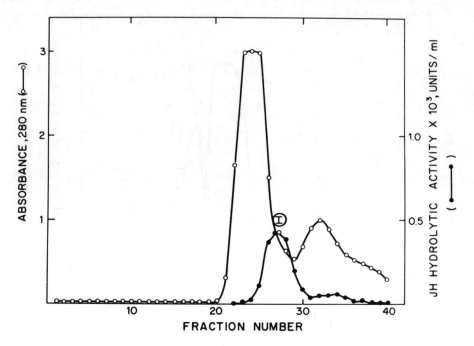

Figure 9. Gel Permeation Chromatography of Fourth Instar *M. sexta* Hemolymph (pH 7.3, 4°, 0.1 M NaCl, Sephadex G-100).

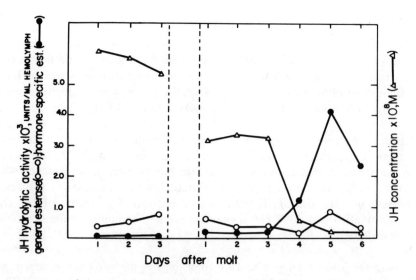

Figure 10. Esterase Levels and Juvenile Hormone Concentration in the Hemolymph of *M. sexta* During Late Larval Development.

esterase remains very low until the fourth day of the fifth instar, when it increases more than 30-fold. It remains at a high level for a short period and then declines.

In *Galleria mellonella*, injected hormone will not prevent metamorphosis if administered late in the fifth larval instar. It has been postulated that only hormone present during a critical period will prevent pupation (Sehnal and Meyer, 1968), and after that period, no amount of hormone will be effective. Thus it appears that for normal metamorphosis to proceed, the hormone titer must fall below the critical level for a short period. Similar results have been obtained with *H. cecropia* (Riddiford, 1972). *Bombyx mori* (Kimura, 1974) and *M. sexta* (Truman *et al.*, 1974). In the latter animal, we postulate that the juvenile hormone-specific esterases, appearing on the fourth day of the fifth instar, rapidly destroy all forms of the hormone and lower the titer below this critical level for commitment to metamorphosis. Approximate calculations show that the juvenile hormone-specific esterase levels reported in Figure 10 are indeed sufficient to account for a considerable decrease of the hormone concentration in the hemolymph. In these calculations we used for the amount of hormone, H, per insect in early fifth instar *M. sexta* larvae the value reported by Weirich and Wren, 1973 (3×10^{-11} moles), and for the rate of hormone production by the *corpora allata* (V_p) the value found in the adult female *M. sexta* (approximately 5.6×10^{-15} moles sec^{-1} per pair of *corpora allata*) (Judy *et al.*, 1973; Metzler *et al.*, 1972).

Juvenile hormone is removed from hemolymph through hydrolysis by general and hormone-specific esterases or through uptake by various tissues, and all of these processes appear to be first order with respect to the hormone (Sanburg *et al.*, 1975a). The steady state hormonal level in the hemolymph, then, requires that

$$V_p = k_d H$$

where k_d is the rate constant for the disappearance of the hormone. This constant is the sum of the rate constants of the three processes outlined above:

$$k_d = \frac{k_G[G]}{1 + \dfrac{B}{K}} + k_E[E] + \frac{k_s}{1 + \dfrac{K}{B}}$$

where $k_G[G]$ is the apparent rate constant of the hydrolysis of juvenile hormone by general esterases, $k_E[E]$ is that of the hydrolysis of the hormone by hormone-specific esterases and k_s is the apparent rate constant for the uptake of juvenile hormone by the tissues. Because of the presence of the carrier protein at concentration B and dissociation constant K and because general esterases hydrolyze only unbound hormone, the first term on the right side of the equation above must be corrected for the amount of juvenile hormone bound. In the same manner, if only bound hormone is taken up by tissues (Sanburg *et al.*, 1975b) the third term must also be corrected. In early fifth instar larvae $k_E[E]$ has the value of 1.4×10^{-4} sec^{-1} and the corrected general esterase rate constant:

$$\frac{k_G[G]}{1 + \dfrac{B}{K}} \; ,$$

is equal to 0.11×10^{-4} sec^{-1}. From the estimated values of V_p, H, $k_E[E]$, $k_G[G]$, B, and K one can calculate a value for k_s of 0.4×10^{-4} sec^{-1} for early fifth instar larvae. Thus during the first half of the fifth instar, approximately 75% of the hormone is degraded by hormone-specific esterases and only about 20% reaches the various tissues. This would indicate that even in this stage the control of the juvenile hormone titer in hemolymph is determined primarily by low levels of hormone-specific esterases. Preliminary experiments, determining the fate of labeled hormone-carrier protein complex injected into larvae, support this conclusion. Then, assuming the same constant values for V_p and k_s throughout the larval life, we used the data on hormone-specific esterase levels reported in Figure 10 to calculate the juvenile hormone concentration in the hemolymph at various times during the instars. These hormone levels are also shown in Figure 10. On the fifth day of the fifth instar the increase in juvenile hormone-specific esterase titer is indeed sufficient to lower the hormone concentration in the hemolymph from 3.3×10^{-8} M to 0.16×10^{-8} M, a 20-fold decrease. A change in concentration of this magnitude compares favorably with those measured for steroid hormones in the mammalian circulatory system (Nocenti, 1968b). This comparison further reinforces the view that juvenile hormone-specific esterases are capable of causing biologically significant changes in juvenile hormone titers. Again, experiments with labeled hormone complexed to carrier protein and injected into M. sexta larvae verify this prediction. The average value of $k_G[G]$ is 4×10^{-4} sec^{-1}, but after correction for the presence of carrier protein it is 0.15×10^{-4} sec^{-1}. During the first half of the fifth instar the uncorrected value is approximately three times larger than the rate constant due to hormone-specific esterases and tissue uptake. Thus, in the absence of carrier protein a large majority of the juvenile hormone molecules would be degraded by the general esterases, and control of hormone titers by other means would be impossible. However, with a physiological concentration of carrier protein of 8×10^{-6} M, the action of the general esterases is negligible and the controlling factor becomes the juvenile hormone-specific esterases. The calculations outlined above were based on data obtained for the production of juvenile hormone by *corpora allata* from adult female M. sexta. The concentration of hormone-specific esterases is so high, even in early fifth instar larvae, that the rate of hormone production in larvae must be at least as high in order to account for the experimentally observed juvenile hormone concentration. Larval and adult levels of hormone production have been estimated to be similar in Cecropia (Williams, 1961). Although evidence has been presented that production decreases during late larval life (Williams, 1961), we feel that the control of juvenile hormone levels in the hemolymph at the time of metamorphosis could well be determined primarily by changes in esterase levels.

Using the data concerning the rates of production and degradation, some conclusions can also be derived about the spatial distribution of juvenile hormone in the hemolymph. Employing a simplistic model of closed circuit hemolymph in a steady state, one can show that the hormone concentration ([H]) at a distance x from the *corpora allata* is expressed by the equation

$$[H]_x = \frac{V_p T}{v(1-e^{-k_d T})} \quad e^{(-\frac{k_d T}{L} \quad x)}$$

where T is the average circulation turnover time, L is the average length of the hemolymph circuit and v is the total volume of the circulating hemolymph. Based on the estimated values, v = 1 ml, T = 1000 sec and L = 10 cm, we find that on the second day of the fifth instar the values for $[H]_{x=0}$ and $[H]_{x=L}$ are 3.3 x 10^{-8} M and 2.7 x 10^{-8} M, respectively. On the fifth day of the fifth instar, the same calculations yield 0.6 x 10^{-8} M and 0.016 x 10^{-8} M respectively for the same terms. Thus, according to this model, in the early fifth instar there is virtually no concentration gradient of hormone in the hemolymph; the same high concentration of juvenile hormone is maintained in all parts of the insect and most of the hormone molecules recirculate several times before removal. A completely different picture emerges when high hormone-specific esterase levels prevail in the hemolymph. At the distal parts of the insect (x=L) a 170-fold decrease in the hormone concentration is calculated with respect to the concentration on the second day. The decrease in concentration close to the *corpora allata* (x=0) is 6-fold, *i.e.* a 35-fold gradient exists along the length of the circulatory path on the fifth day. Thus degradative enzymes of circulating hormones affect much more the distal than the proximal parts of the insect. Such a model would be eminently suitable for rationalizing slow "waves of differentiation" in organisms with a rapidly circulating carrier fluid.

Catabolic enzymes thus play a key role in regulating hormone titers in the hemolymph. Whether the same or similar hormone-specific esterases are also induced in the fat body and other tissues of the fifth instar larva and whether binding proteins can protect against other degradative enzymes, such as epoxide hydrases, remains to be explored.

Juvenile Hormone at Target Tissues

A recent report from the laboratory of Schmialek (1973) suggests that epidermal tissue from *Tenebrio molitor* contains a macromolecule with high affinity for juvenile hormone (K = 10^{10} M). This "receptor," which is reported to be a nucleoprotein, is detected by incubation of the tissue with labeled juvenile hormone, followed by extraction with a buffer containing a detergent, Triton X-100. The resulting solution is subjected to centrifugation and gel filtration on agarose. Labeled hormone emerges from the column in the macromolecular region with the nucleoprotein fraction. The "receptor" activity was destroyed by treatment with proteolytic enzymes.

We have used the techniques of Schmialek *et al.* (1973) to investigate the interactions of labeled juvenile hormone with fourth or fifth larval instar imaginal disks from *M. sexta.* We obtained two labeled macromolecular fractions that could be separated by gel filtration. The larger was shown to contain intact hormone and the smaller, the free acid. However, controls showed that the larger fraction was also obtained when labeled hormone was simply dissolved in the Triton X-100 buffer, and we appear to be dealing with nothing more than a mixed micelle of hormone and Triton. We believe that the smaller "macromolecule" is the Triton-free acid micelle. The particularly deceiving feature of these artifacts is their behavior on gel filtration and acrylamide gels. They emerge in sharp symmetrical peaks or bands and can be rechromatographed with no change in their behavior.

It became clear to us that hornworm imaginal disks had a great capacity for destruction of the hormone, which is in agreement with reports concerning *Drosophila* disks (Chihara *et al.,* 1972). We attempted to reduce this by incubation of disks with hormone-carrier protein complex, but under these conditions so little hormone entered the disk tissue that it was not possible to follow it. We found that incubation of the disks in Grace's insect tissue culture medium containing 10^{-3} M DFP overnight at 0° did not destroy the disks or impair their ability to take up hormone from the medium. After several attempts we were able to observe a small incorporation into a membrane fraction which could be isolated by isopycnic centrifugation. We found this material also in larval abdominal muscle, but not in midgut. Whether or not this membrane-bound hormone has any significance in the action of the hormone remains to be determined.

Thus, the events occurring at the target tissues are not yet understood, and much remains to be done. An elaborate hypothesis for hormone action in target tissue has been advanced, without any experimental support (Williams and Kafatos, 1972). A number of studies by the Ilans (Ilan *et al.,* 1972) have centered on the role of juvenile hormone in translational control of protein synthesis. Some evidence for a specific transfer RNA, related to the presence of hormone, has been advanced. It has been shown that juvenile hormone is only effective on cells which have not completed a premolting round of DNA synthesis (Schneiderman, 1972).

It seems likely that results obtained from the rapid pace of research on juvenile hormones will soon illuminate many of the unanswered questions. It is an area that offers the excitement of understanding the invertebrate hormone systems that so often vary from the familiar mammalian cases, and the opportunity to engage in pragmatic research that will lead to new specific weaponry in the battle with the insects.

Acknowledgements

We wish to acknowledge the technical assistance of Ms. H. Seballos, Mr. J. Ong and Mr. D. Dill. We would like to thank Dr. Judith H. Willis for reading and commenting on the manuscript. This research was supported in part by National Science Foundation Grant GB 8436 A1 and National Institutes of General Medical Sciences Grant GM 13863 to JHL.

Note Added in Proof

Since this review was written, many papers relevant to the proposed ideas have appeared, from which we call attention to three. The Zoecon group (Jennings *et al*, 1975. *Chem. Comm.* *21*) has now demonstrated the conversion of isotopically labeled homomevalonate to methyl homojuvenate by intact *corpora allata.* A specific binding protein for juvenile hormone, much like the one we have reported from *M. sexta*, has been found in the hemolymph of *Plodia interpunctella* (Ferkovich *et al.*, 1975. *Insect Biochem. 5*, 141). A paper by Nijhout and Williams (1974. *J. Exp. Biol. 61*, 493) reports juvenile hormone levels in the hemolymph of larval *M. sexta* using bioassay techniques. Although these authors attribute the 23-fold decrease in JH levels during the fifth instar to cessation of *corpus allatum* function, the experimental points fit nicely on the theoretical curve we present in Figure 10, based upon esterase activity as the determinant of hormone levels, and with the assumption that hormone production does not cease.

REFERENCES

Ajami, A.M. and Riddiford, L.M. (1973). *J. Insect Physiol. 19*, 635.

Akamatsu, Y. and Law, J.H. (1970). *J. Biol. Chem. 245*, 709.

Bowers, W.S., Thompson, J.J. and Uebel, E.C. (1965). *Life Sci. 4*, 2323.

Brockman, H.L., Law, J.H. and Kézdy, F.J. (1973). *J. Biol. Chem. 248*, 4965.

Chihara, C.J., Petri, W.H., Fristrom, J.W. and King, D.S. (1972). *J. Insect Physiol. 18*, 1115.

Chino, H., Murakami, S. and Harashima, K. (1969). *Biochim. Biophys. Acta 176*, 1.

Doane, W.W. (1973) in *Developmental Systems: Insects 2*, eds. Counce, S.J. and Waddington, C.H. (New York: Academic Press), p. 291.

Emmerich, H. and Hartmann, R. (1973). *J. Insect Physiol. 19*, 1663.

Gilbert, L.I. and King, D.S. (1973) in *The Physiology of Insecta* 2nd ed. 1, ed. Rockstein, M. (New York: Academic Press), p. 249.

Hall, Z.W., Bownds, M.D. and Kravitz, E.A. (1970). *J. Cell Biol. 46*, 290.

Ilan, J., Ilan, J. and Patel, N.G. (1972) in *Insect Juvenile Hormones*, eds. Menn, J.J. and Beroza, M. (New York: Academic Press), p. 43.

Judy, K.J., Schooley, D.A., Dunham, L.L., Hall, M.S., Bergot, B.J. and Siddall, J.B. (1973). *Proc. Nat. Acad. Sci. USA 70*, 1509.

Kanai, M., Raz, A. and Goodman, DeW.S. (1968). *J. Clin. Invest. 47*, 2025.

Katzenellenbogen, J.A., Johnson, H.J.Jr. and Myers, H.N. (1973). *Biochemistry 12,* 4085.

Kimura, S. (1974). *J. Insect Physiol. 20,* 887.

Kramer, K.J., Sanburg, L.L., Kézdy, F.J. and Law, J.H. (1974). *Proc. Nat. Acad. Sci. USA 71,* 493.

Lederer, E. (1969). *Quart. Rev. Chem. Soc. 23,* 453.

Metzler, M., Dahm, K.H., Meyer, D. and Röller, H. (1971). *Z. Naturforsch. 26,* 1270.

Metzler, M., Meyer, D., Dahm, K.H., Röller, H. and Siddall, J.B. (1972). *Z. Naturforsch. 27,* 321.

Meyer, A.S., Schneiderman, H.A., Hanzmann, E. and Ko, J.H. (1968). *Proc. Nat. Acad. Sci. USA 60,* 853.

Nocenti, M.R.(1968a) in *Medical Physiology* 12th ed. 2, ed. Mountcastle, V.B. (Saint Louis, Missouri: C.V. Mosby Co.), p. 962.

Nocenti, M.R.(1968b) in *Medical Physiology* 12th ed. 2, ed. Mountcastle, V.B. (Saint Louis, Missouri: C.V. Mosby Co.), p. 992.

Popják, G. (1969) in *Methods in Enzymology* 15, ed. Clayton, R.B. (New York: Academic Press), p. 393.

Pratt, G.E. and Tobe, S.S. (1974). *Life Sci. 14,* 575.

Reibstein, D. and Law, J.H. (1973). *Biochem. Biophys. Research Comm. 55,* 266.

Riddiford, L.M. (1972). *Biol. Bull. 142,* 310.

Riddiford, L.M. and Ajami, A.M. (1973). *J. Insect Physiol. 19,* 749.

Rodwell, V.W., McNamara, D.J. and Shapiro, D.J. (1973) in *Advances in Enzymology* 39, ed. Meister, A. (New York: John Wiley and Sons), p. 373.

Röller, H., Dahm, K.H., Sweeley, C.C. and Trost, B.M. (1967). *Angew. Chem. Int. Ed. Engl. 6,* 179.

Sanburg, L.L., Kramer, K.J., Kézdy, F.J. and Law, J.H. (1975a). *J. Insect Physiol. 21,* 873.

Sanburg, L.L., Kramer, K.J., Kézdy, F.J., Law, J.H. and Oberlander, H. (1975b). *Nature 253,* 266.

Schmialek, P. (1973). *Nature 245,* 267.

Schooley, D.A., Judy, K.J., Bergot, B.J., Hall, M.S. and Siddall, J.B. (1973). *Proc. Nat. Acad. Sci. USA 70*, 2921.

Schmailek, P., Borowski, M., Geyer, A., Miosga, V., Nundel, M., Rosenberg, E. and Zapf, B. (1973). *Z. Naturforsch. 28*, 453.

Schneiderman, H.A. (1972) in *Insect Juvenile Hormones*, eds. Menn. J.J. and Beroza, M. (New York: Academic Press), p. 3.

Sehnal, F. and Meyer, A.S. (1968). *Science 159*, 981.

Siddall, J.B., Anderson, R.J. and Henrick, C.A. (1971). *Proc. 23rd Int. Congr. Pure Appl. Chem. 3*, 17.

Slade, M. and Zibitt, C.H. (1972) in *Insect Juvenile Hormones*, eds. Menn, J.J. and Beroza, M. (New York: Academic Press), p. 155.

Sternberg, J., Vinson, E.B. and Kearns, C.W. (1953). *J. Econ. Entomol. 46*, 513.

Trautmann, K.H. (1972). *Z. Naturforsch. 27B*, 263.

Truman, J.W., Riddiford, L.M. and Safranek, L. (1974). *Devel. Biol. 39*, 247.

Weirich, G. and Wren, J. (1973). *Life Sci. 13*, 213.

White, A.F. (1972). *Life Sci. 11*, 201.

Whitmore, D.Jr., Whitmore, E. and Gilbert, L.I. (1972). *Proc. Nat. Acad. Sci. USA 69*, 1592.

Whitmore, E. and Gilbert, L.I. (1972). *J. Insect Physiol. 18*, 1153.

Williams, C.M. (1956). *Nature 178*, 212.

Williams, C.M. (1961). *Biol. Bull. 121*, 572.

Williams, C.M. and Kafatos, F.C. (1972) in *Insect Juvenile Hormones*, eds. Menn, J.J. and Beroza, M. (New York: Academic Press), p. 29.

Wyatt, G.R. (1972) in *Biochemical Actions of Hormones* 3, ed. Litwack, G. (New York: Academic Press), p. 385.

STRUCTURAL AND FUNCTIONAL ASPECTS OF THE PROTEIN SYNTHESIZING APPARATUS IN THE ROUGH ENDOPLASMIC RETICULUM

*David D. Sabatini, George Ojakian, Mauricio A. Lande, John Lewis, Winnie Mok, Milton Adesnik, Gert Kreibich

Department of Cell Biology
New York University School of Medicine
550 First Avenue
New York, New York 10016

Introduction

The successful implementation of a genetic program requires that upon completion of translation polypeptides released from ribosomes are transferred to their final destination in the cell. This appears to be a rather formidable process since, in a eukaryotic cell, the fate of a protein can be at least as diverse as the membranous structures within the cell and the subcellular compartments which they separate. For many proteins, these subcellular compartments serve only as the initial destination; transfer to another compartment or export from the cell may be their ultimate fate.

From observations made mainly on secretory cells (Palade, 1958; Siekevitz and Palade, 1960; Redman and Sabatini, 1966) it is now apparent that the fate of proteins is related to their site of synthesis. In the cytoplasm of most eukaryotic cells, there are at least two types of polysomes which are functionally specialized according to the fate of their products: those which are free in the cell sap and discharge their polypeptides into the soluble phase of the cytoplasm and those which are bound to the membranes of the endoplasmic reticulum (ER) and discharge their products vectorially, either into these membranes or across them into the ER lumen. Recently, the existence of a third fraction of cytoplasmic ribosomes has been reported in rapidly growing yeast cells in which polysomes are bound to the outer surface of the mitochondria (Kellems and Butow, 1972;

---

*Symposium participant.

Kellems *et al.,* 1974; Kellems and Butow, 1974). These ribosomes are also likely to discharge their products vectorially into, or across, the mitochondrial membranes into the matrix space. Evidence is also accumulating which suggests a somewhat similar situation within mitochondria (Kuriyama and Luck, 1973) and chloroplasts (Chua *et al.,* 1973; Margulies and Michaels, 1974). Ribosomes have been found to exist within these organelles either free or attached to the inner membranes.

An understanding of the operation of the selective mechanisms which serve to insure the topographic segregation of ribosomes translating specific kinds of mRNA appears to be necessary to fully comprehend how the subcellular fate of proteins is determined. The nature of these mechanisms, however, is as yet obscure. It is not clear, for example, if information which directs the binding of ribosomes to membranes is translated as part of the polypeptide chain (Blobel and Sabatini, 1971a; Sabatini *et al.,* 1972; Brownlee *et al.,* 1972; Milstein *et al.,* 1972; Harrison *et al.,* 1974a, b), or messengers destined for translation on ER membranes are segregated within the cell by mechanisms which operate through the recognition of untranslated messenger sequences.

Structural Aspects of Binding of Ribosomes to Endoplasmic Reticulum Membranes

Rat liver rough microsomes (RM) are membrane vesicles formed during tissue homogenization by fragmentation of the rough endoplasmic reticulum cisternae. Ribosomes remain bound to the outer surface of rough microsomes and part of the cisternal content of the ER is retained in their luminal cavities (Palade and Siekevitz, 1956; Kreibich *et al.,* 1973, 1974). Even after extensive purification rough microsomes are active in protein synthesis *in vitro* and capable of very effectively coupling puromycin into the nascent polypeptides contained in the attached ribosomes. Most polypeptides terminated *in vitro* and peptidyl-puromycin molecules released from the attached ribosomes are vectorially discharged into the vesicular cavities which represent the lumen of the ER cisternae (Redman, Siekevitz and Palade, 1966; Redman and Sabatini, 1966; Kreibich and Sabatini, 1973, 1974).

Initial attempts to remove bound ribosomes from membranes by adding chelating agents such as EDTA to rough microsomes suggested that the large ribosomal subunits bear stronger sites of attachment to the membranes. Low concentrations of EDTA added to microsomes suspended in solutions of low ionic strength led to the preferential release of unfolded small ribosomal subunits from the membranes (Sabatini *et al.,* 1966), while large subunits of bound active ribosomes (containing nascent polypeptides) were not as easily released. Electron microscopic studies carried out with fixed cells and with isolated microsomes demonstrated that the cleft or partition between subunits is parallel to the ER surface and also established that binding to the membranes occurs only through the large ribosomal subunits (Sabatini *et al.,* 1966; Shelton and Kuff, 1966; Florendo, 1969).

The phenomenon of vectorial discharge of secretory polypeptides and the finding that large subunits of active ribosomes are more tightly associated to the microsomal membranes than large subunits of ribosomes which incorporate low levels of amino acid radioactivity *in vivo,* suggested that nascent polypeptides of bound ribosomes are directly and intimately associated with the underlying ER membranes. This was demonstrated by the observation (Sabatini and Blobel, 1970) that mild proteolysis of rough microsomes, which caused extensive removal of proteins and ribosomes from the outer face of the ER membranes, did not lead to the complete digestion of nascent polypeptides. These polypeptides were cleaved into two segments both of which were largely protected from proteolysis. One of the segments remained intra-ribosomal and was removed together with the detached ribosomes, while the other, corresponding to the extra-ribosomal piece of the nascent chain which contains the amino terminal segments, was retained in the microsomes. Extra-ribosomal fragments left with the vesicles denuded of ribosomes were inaccessible to the added proteases unless detergents were added to dissolve the proteolyzed membranes.

Through the use of puromycin to release nascent polypeptides in media of high ionic strength containing Mg^{2+} concentrations sufficiently high to preserve ribosomal structure (Blobel and Sabatini, 1971b; Nonomura *et al.,* 1971; Sabatini *et al.,* 1971), a role of the nascent polypeptides in maintaining the association between large ribosomal subunits and membranes of rough microsomes was firmly established (Adelman *et al.,* 1973). Under these conditions puromycin produces the *in vitro* dissociation of isolated free or bound polysomes into 40S and 60S undenatured ribosomal subunits.

Puromycin treatment $(10^{-3} - 10^{-4}M)$ of rough microsomes in a solution of high ionic strength containing Mg^{2+} (500-700 mM KC1, 50 mM Tris-Hc1, pH 7.4, 2.5 - 5 mM Mg^{2+}) led to the release of most ribosomes from the membranes as functionally viable subunits (Adelman *et al.,* 1973a). Although the reaction coupling puromycin to nascent polypeptides proceeds effectively at low ionic strengths (25-100 mM KC1), rapid and extensive puromycin dependent release of subunits from the membranes occurred only when the salt concentration was raised to values higher than 100 mM KC1. These observations and the finding that inactive ribosomes lacking nascent polypeptides, as well as active ribosomes containing very short nascent polypeptides, are released from microsomes in high ionic strength solutions containing Mg^{2+} without added puromycin, led to the conclusion that at least two types of interactions mediate the attachment of ribosomes to microsomal membranes. The first is sensitive to ionic strength and probably involves electrostatic interactions between specific sites on the large ribosomal subunits and receptor sites for ribosomes on the microsomal membranes. The other interaction, which is puromycin sensitive, exists only in "active" ribosomes and is mediated by nascent polypeptide chains sufficiently long to anchor the ribosomes to the membranes. As the growing nascent chain traverses the membrane and protrudes into the lumen of the microsomal vesicle, tertiary folding of the polypeptide would provide the "anchor" necessary for firm ribosomal attachment.

One may hypothesize that the first interaction between the extraribosomal segment of a nascent polypeptide chain and the microsomal membrane results from hydrophobic bonding between membrane components and the amino terminal segment of the chain with the pertinent binding information being contained in non-polar sequences or modifications. The interaction of nascent chains with membrane-bound enzymatic systems involved in modifying the polypeptides by processes such as hydroxylation (Lazarides *et al.,* 1970; Redman and Cherian, 1972) or crosslinking (DeLorenzo *et al.,* 1966) may play an additional role in reinforcing ribosome binding.

The existence of a subunit dissociation-reassociation cycle during protein synthesis in bound ribosomes has been demonstrated by utilizing an *in vitro* protein synthesizing system capable of partial chain termination and puromycin as an artificial terminator of polypeptide growth (Borgese *et al.,* 1972). It was found that upon *in vitro* termination of nascent chains, small subunits of bound ribosomes easily exchanged with free small subunits added to rough microsomes. This exchange was readily measured by uptake or release of labeled subunits from microsomes. On the other hand, a similar exchange between free and bound large subunits could not be observed, although exchange of large subunits was easily demonstrated with free polysomes. When added large subunits were present in a system containing rough microsomes and termination was induced, the added large subunits combined with and effectively removed from the microsomes the majority of the bound small subunits.

Purified free and bound large subunits are both capable of binding effectively to receptor sites on rough microsomal membranes stripped of ribosomes (see following section). One must therefore conclude that the lack of exchangeability of bound large subunits does not reflect an incompetence of the added subunits, but the fact that large subunits are not released from rough microsomes under the *in vitro* conditions of termination employed. This is supported further by the fact that, in highly diluted suspensions of rough microsomes, maintained at normal ionic strength without added free subunits, puromycin was able to induce a release of bound small subunits. Under these conditions no large subunits were released from the microsomes by puromycin, as would have been expected if a ribosomal dissociation-association reaction with the membranes occurred. It should be stressed however, that lack of exchange, or release, of subunits *in vitro* may only reflect an inadequacy of the microsomal system to reproduce *in vivo* conditions such as the presence of factors which may serve to regulate the exchange process. Recently however, evidence has been presented supporting a model of assembly of bound polysomes in which large subunits bind directly to the membranes in the absence of protein synthesis and therefore can remain associated with them throughout several rounds of translation (Baglioni *et al.,* 1971).

Binding Sites For Ribosomes in Microsomal Membranes

There is now considerable evidence for the existence of ribosome binding sites in microsomal membranes derived from rough endoplasmic reticulum (Suss *et al.,* 1966; James *et al.,* 1969; Williams and Rabin, 1969, 1971; Roobol and Rabin, 1971; Scott-Burden and Hawtrey, 1971, 1973; Shires *et al.,* 1971a, b. 1973; Sunshine *et al.,* 1971;

Hochberg *et al.,* 1972; Nolan and Minro, 1972; Burke and Redman, 1973; Ekren *et al.,* 1973; Jothy *et al.,* 1973; Pitot and Shires, 1973; Shires and Pitot, 1973; Borgese *et al.,* 1972; Rolleston, 1972; Rolleston and Mak, 1973; Rolleston and Lamm, 1974). The binding sites involved and the binding reaction have been characterized using membranes of rat liver rough microsomes stripped of ribosomes by the puromycin-high salt procedure (Borgese *et al.,* 1974). For the binding assay, labeled ribosomes and ribosomal subunits were prepared by the puromycin-KC1 procedure from animals which had received [^{3}H] orotic acid a precursor of RNA. At low ionic strength and 0°, the stripped rough microsomes were shown to have a high capacity to bind ribosomes with saturation of the binding sites occurring at ribosome levels similar to those found in native rough microsomes (Table I). Nonstripped rough microsomes, which have their sites occupied by native ribosomes, showed almost no binding capacity for ribosomes at 0° (Table I). The ribosome binding capacity of smooth membranes treated with puromycin-high salt was significantly lower than the capacity of stripped rough microsomes, a result which suggests that the smooth character of the membranes does not depend primarily on introcellular segregation of ribosomes in other regions of the cell making them unavailable for

TABLE I

BINDING OF ^{3}H-80S RIBOSOMES TO VARIOUS MEMBRANE FRACTIONS FROM RAT LIVER AT LOW IONIC STRENGTH

Tritiated 80S ribosomes were prepared by the puromycin (10^{-3}M)-high salt procedure (500 mM KC1, 50 mM Tris/HC1, pH 7.6, 5 mM MgCl$_2$; Adelman *et al.,* 1973a) from rough microsomes (RM) which were isolated 40 hr after an i.p. *in vivo* injection of [^{3}H] orotic acid (spec. activity 17.8 Ci/mmole, 1.6 μC were injected per gram rat). Binding of radioactive 80S ribosomes at saturating concentrations (160μg 80S per 80μl) was measured by a flotation assay (Borgese *et al.,* 1974) after incubating membranes at the concentrations indicated in TKM at 0° for 30 min. The Km was computed from Scatchard plots. Higher concentrations of mitochondrial protein were used to account for the unexposed inner membranes and matrix.

Membrane Fraction	mg Membrane Protein in 80μ Sample	Ribosomes Bound μg	Km(x10^7M^{-1})
Stripped RM	0.1	24.2	7.1
Stripped SM	0.17	15.0	5.7
SM	0.16	11.0	5.2
RM	0.2	6.7	0.81
Golgi	0.28	6.1	0.92
Mitochondria	0.6	7.8	0.80
Plasma Membrane	0.2	2.2	0.92

binding. Erythrocyte membranes, mitochondrial membranes, plasma membranes, or membranes derived from the Golgi apparatus were only able to bind ribosomes at very low capacities or were unable to bind ribosomes at all (Table I). The affinity constants (5-$8 \times 10^7 M^{-1}$) of the binding reactions for both stripped rough and smooth microsomes, as determined by Scatchard plots, were found to be similar (Table I) although the binding capacity of the smooth membranes was considerably lower suggesting that the same type of site is involved in both membranes. The fact that treatment of smooth microsomes under conditions for stripping increases their binding capacity (Table I) can be explained by the presence of some membrane-bound ribosomes in this fraction, which contains ∿20% of the RNA of rough microsomes. The binding sites of ER membranes were easily destroyed *in vitro* by heat treatment or by mild proteolytic digestion which produced only minor changes in the protein electrophoretic patterns obtained in SDS acrylamide gels. Since enzymatic treatment of the membranes with phospholipase C or nucleases failed to affect their binding ability for ribosomes, it is likely that proteins are the primary components of ribosome binding sites in ER membranes.

At low ionic strengths, the binding was stronger but the reaction showed a relatively poor specificity for ribosomes. Monomers prepared from free or bound polysomes, as well as large or small subunits, bound almost equally well to stripped rough membranes. Free ribosomes prepared from reticulocytes were also able to bind, but E. coli ribosomes were not accepted by ER membranes (Table II). At higher ionic strengths, however, a selectivity for the binding of large subunits became apparent. Between 100-150 mM KC1, using equimolar inputs of both large and small subunits, almost four times as many large subunits were bound to the membranes as small. Similar binding constants and a preference for large over small subunits has been demonstrated by Rolleston (1972) and Rolleston and Lamm (1974), using a mouse liver assay system.

The finding that large subunits can be preferentially bound to ER membranes under nearly physiological ionic strengths is compatible with models of assembly of bound polysomes (Fig. 10b) in which large subunits are membrane-bound before they join the initiation complex. This model also agrees with the observations of Baglioni *et al.* (1971) that, in tissue culture cells, newly synthesized large subunits—but not small subunits— appear in a membrane-bound form soon after emerging from the nucleus. The same finding also raises the possibility shown in Figure 10b, that after termination, bound large subunits may not necessarily detach from the membranes. If this is the case it is unlikely that a recognition mechanism involving the nascent chain alone determines that translation of certain messengers occurs in the bound condition. A nascent-chain signal could lead to either reinforcement of the attachment or rejection of bound subunits from the binding sites (Harrison *et al.*, 1974a, b). Selection could also be imposed by specific initiation complexes which may be involved in the recognition of bound large subunits; in several cases initiation factors have been shown to regulate the affinity of subunits for mRNA (Groner *et al.*, 1972; Lee-Huang and Ochoa, 1972; Nudel *et al.*, 1972).

Ribosomal binding sites for the microsomal membranes have not yet been well characterized. Aurintricarboxylic acid (ATA), an inhibitor of initiation, has been shown

TABLE II

BINDING OF LARGE SUBUNITS FROM RAT LIVER
AND E. COLI TO RM*str* OF RAT LIVER

Stripped rough microsomes (0.37 mg) were incubated in 80μl of low salt buffer (100 mM KC1, 50 mM Tris/HC1, pH 7.6, 5 mM MgC1$_2$) for 30 min at 0° with 0.11 mg ^{3}H-60S or ^{14}C-50S subunits. ^{3}H-60S subunits were collected from linear sucrose gradients (10 - 25% sucrose - HSB; 165 min, 35,000g SW 41, 20°) after stripping of [^{3}H] orotic labeled rough microsomes (see legend to Table I). ^{14}C-50S subunits were obtained as a gift from Dr. W. Szer (Szer and Leffler, 1974).

Amounts of subunits bound are expressed in ρmol, calculated on the basis of particle weights of 3.0 x 10^6 for 60S and 1.8 x 10^6 for 50S.

Ribosomal Subunits	Ribosomes Bound (ρmol)	KM(x10^{7}M^{-1})
60S	5.1	8.7
50S (crude)	0.65	0.17
50S (washed with 1M NH$_4$C1)	0.75	0.17

to inhibit binding (Borgese *et al.*, 1974). This dye acts directly on ribosomes, as shown by the fact that ribosomes incubated with $\sim$10^{-4}M ATA and then recovered by centrifugation have a considerably lower affinity for membranes. Since mild proteolytic digestion of the ribosomes also severely affects their ability to bind to the membranes, proteins are probably involved in the binding reaction.

Taking advantage of the *in vitro* affinity of ribosomes for membranes, attempts have been made to reconstitute rough microsomes using stripped microsomal membranes and polysomes containing nascent chains (Shires *et al.*, 1973; Shires and Pitot, 1973; Burke and Redman, 1973; Borgese *et al.*, 1974). In our view, however, these attempts have failed since polypeptides discharged by puromycin from the reattached ribosomes remained on the outer surface of the vesicles and were not vectorially discharged into the cavities as could be shown by their accessibility to added proteases (Borgese *et al.*, 1974). The functionality of the mechanism for vectorial transfer is likely to be insured by several recognition mechanisms which operate during the assembly of bound polysomes and coordinate the binding of large subunits to membranes with the growth of a specific type of protein chain.

Previous analysis by SDS-polyacrylamide gel electrophoresis of the proteins present in rat liver free and membrane-bound ribosomes suggested that at least one extra protein (MW∿50,000) is present in the large subunits of free polysomes (Borgese *et al.*, 1973). Similar differences have also been found between free and bound ribosomes in bacteria (Brown and Abrams, 1970) and in chicken embryos (Fridlender and Wettstein, 1970). These observations led us to consider the possibility that additional components of free large subunits may play a regulatory role as factors which prevent ribosome binding or promote ribosome detachment.

More recent studies using both one-dimensional SDS polyacrylamide gel electrophoresis and two-dimensional electrophoresis in gradient gels have revealed several other differences in the protein components of rat liver bound and free ribosomal subunits prepared without the use of detergent by dissociation of polysomes in a puromycin-high ionic strength medium (Blobel and Sabatini, 1971b; Adelman *et al.*, 1973a). These differences (Fig. 1) consist of the absence of some bands as well as the appearance of extra bands in both 40S and 60S subunits. Therefore, our current observations suggest that several protein bands in both subunits may correspond to factors involved either in the recognition of specific mRNAs and/or in regulating the binding of selective polysome classes to endoplasmic reticulum membranes.

Translation on Microsomal Membranes

The mechanism of successive decoding involved in the elongation phase of translation requires the relative movement of mRNA with respect to the ribosomes within a polysome. It has not yet been established if, within the cell, large ribosomal subunits of bound polysomes dissociate and reassociate with binding sites in the ER membranes in coordination with the subunit cycle. However, the stability of the large subunit-membrane interaction, demonstrated *in vitro*, and the finding that active ribosomes are linked to ER membranes through their nascent polypeptides, strongly suggest that, within a bound polysome, the association of individual ribosomes with the membrane is maintained throughout the process of elongation. The possibility that ribosomes may move along the messenger by a process involving their cyclic detachment and reattachment to the membranes can therefore be ruled out.

Since successive rounds of translation could lead to large displacements of the messenger RNA in a direction parallel to the membrane plane, efficient messenger utilization by bound ribosomes would be facilitated if: (a) ribosome binding sites were also able to move in the plane of a fluid ER membrane and/or (b) within a domain of the membrane, ribosomes and mRNA remain in a fixed geometrical configuration which insures or facilitates messenger reutilization.

We examined by electron microscopy (Ojakian *et al.*, 1975) the possibility that binding sites bearing ribosomes are mobile in the plane of ER membranes. Such movement would facilitate elongation and also insure the availability of binding sites for large subunits in the vicinity of the newly-formed initiation complexes. Using freeze etching

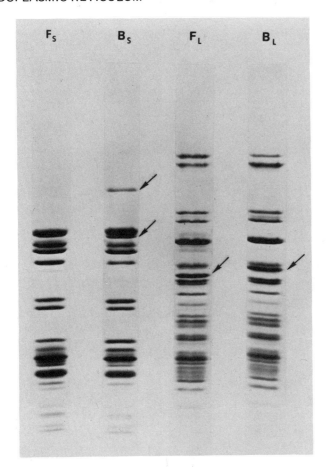

Figure 1. Gradient gel electrophoresis of proteins from rat liver free and membrane-bound ribosomal subunits. Free and membrane-bound ribosomal subunits were prepared as described by Adelman *et al.*, 1973b. Electrophoresis was carried out on a linear 12-18% polyacrylamide slab gel containins SDS which was stained with Coomassie brilliant Blue. Proteins of known molecular weight (ovalbumin, concanavalin A, apoferritin and cytochrome c) were electrophoresed in neighboring tracks. Differences between free and membrane-bound 40S subunits are observed in the 40,000 dalton region while differences between 60S subunits are observed in the 25,000 dalton region.

techniques (*c.f.* Branton and Deamer, 1972), extensive membrane surfaces were visualized by deep-etching rough microsomes which were frozen in distilled water. Bound ribosomes were clearly detected as 300 Å particles on the surface of deep-etched microsomes. Typical polysomal patterns in which ribosomes appeared arranged in rows, circles, semicircles, or small spirals were present in untreated microsomes maintained at $0°$ before fixing as well as in microsomes which were first incubated for 30 minutes at $24°$ to increase membrane fluidity (Figs. 2a and b). Since, in addition to movement of individual

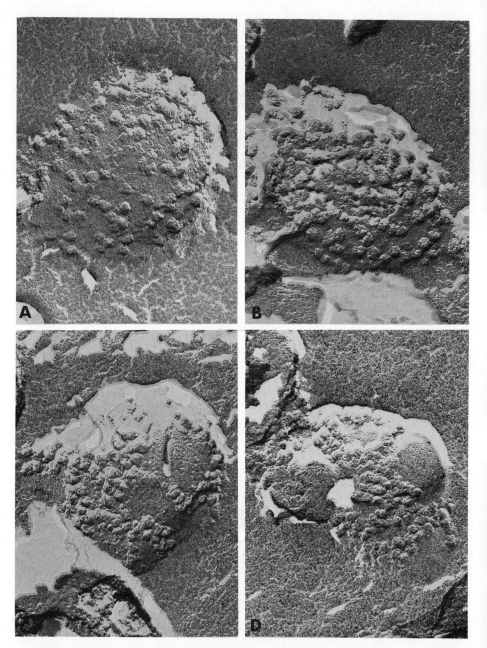

Figure 2. **Freeze etching of rough microsomes showing the clustering of bound ribosomes after ribonuclease treatment.** On the surface of the control deep-etched microsomes (a, b) ribosomes are seen as 300Å particles arranged in rows which correspond to polysomal patterns. After treatment with pancreatic RNase (10μg/ml) at 0° and incubation for 30 min at 24°, ribosomes have moved laterally to form large clusters (c and d), leaving extensive regions of smooth membranes. (162,000X).

ribosomes, whole polysomes could be expected to move as units on the membrane surface, we introduced a mild treatment with RNase to cleave the messenger RNA and allow the random movement of individual ribosomes. This treatment did not lead to detachment of ribosomes from the membranes as determined by analytical sucrose gradients, but instead, ribosomes moved laterally to form large clusters leaving intervening large areas of smooth membrane (Figs. 2c and d). Ribosome movement occurred even at 0°, but clustering was most pronounced in RNase-treated microsomes which were incubated for 30 minutes at 24° after termination of enzymatic digestion. Observations on RNase-digested microsomes fixed and embedded for conventional thin-sectioning electron microscopy (Fig. 3) demonstrated that ribosomes first cluster in small groups, rows and patches followed by subsequent formation of large aggregates or ribosome caps. The sequence of events in ribosomal movement and aggregation is similar in many respects to the capping of antibodies on lymphocyte cell surfaces (Karnovsky et al., 1972). In numerous instances, regions of the microsomal membrane bearing the ribosomal aggregates formed invaginations into the microsomal cavity (Fig. 3b and c), leaving the membrane surface almost completely devoid of ribosomes. This process was apparent also in deep-etched samples in which openings of the invaginations were visible in ribosome-free areas of the microsomal surface. Since bound ribosomes remain firmly attached to the membrane, both during and after ribonuclease digestion, the lateral movement exhibited by these ribosomes can be interpreted as mobility of the receptor or binding site and its attached ribosome as a single unit in a fluid microsomal membrane (Singer and Nicholson, 1972).

Cleavage of messenger RNA followed by ribosome movement would be expected to lead only to randomization of ribosomes, but instead, we observed extensive clustering involving, in many instances, all the ribosomes visible in a microsome profile. This would be expected if RNase treatment, because of cleavage of rRNA on the ribosome surface also increased the adhesiveness of ribosomes for each other.

Although our findings demonstrate that extensive movement of bound ribosomes is possible in vitro, movement of membrane-bound ribosomes within the cell has yet to be demonstrated. In spite of the fact that both rough and smooth microsomes are derived from a continuous system of ER membranes, the demonstration that rough microsomes have considerably more ribosome binding sites than smooth microsomes (see preceeding section) suggests that, in vivo, restrictions are imposed on ribosome receptor mobility. Such restrictions would lead to a segregation of ribosomes in specialized regions and give rise to rough and smooth endoplasmic reticulum.

Relationship of Binding Sites to Other ER Membrane Proteins

The existence of morphological differentiation is characteristic of the endoplasmic reticulum. Rough and smooth microsomes are derived from regions characterized by the presence or absence of ribosomes bound to the membranes. In addition to their ribosome content, rough and smooth cisternae differ extensively in their morphological appearance, e.g. shape, width, degree of binding or stacking of cisternae and their

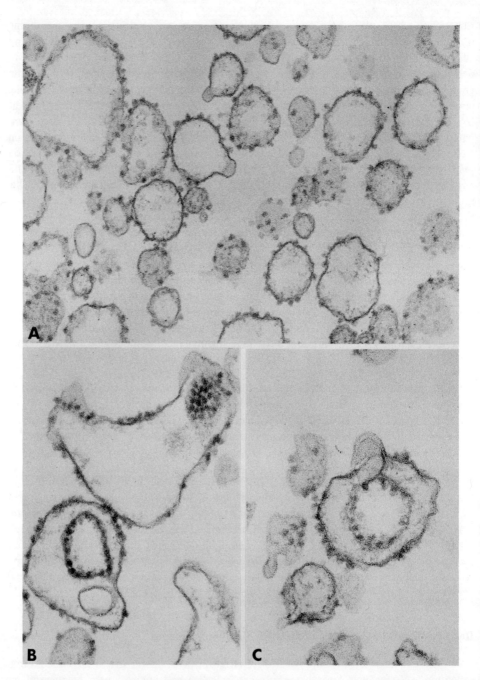

Figure 3.   Electron micrographs of thin-sectioned rough microsomes.   In controls (a) bound ribosomes are distributed homogeneously around the microsome proper.   After RNase treatment followed by incubation at 24°, ribosomes have aggregated on the microsomal membrane surface to form caps or invaginations.   A, (75,000X); B, C, (102,000).

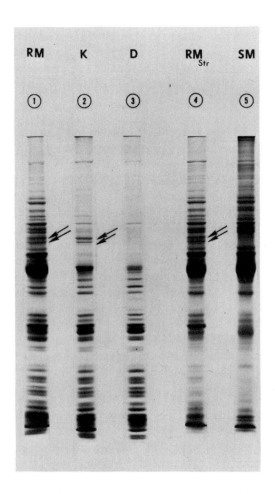

Figure 4. Electrophoretic patterns in SDS acrylamide gels (8-13%) obtained from rough microsomes (RM), stripped rough microsomes (RMstr.), smooth microsomes (SM) and sedimentable subfractions from rough microsomes treated with 2.5 x 10^{-2}M DOC (D) or 2.5 x 10^{-2}M Kyro-EOB (K). Sedimentable subfractions D and K were obtained after centrifugation for 90 min in a #65 Spinco rotor at 40K and 4°. RM and SM contained ∿200µg of protein; other samples were derived from equivalent amounts of microsomes (for details see Kreibich and Sabatini, 1974). Arrows point to peptide bands which are found in RM but not in SM and are part of the network preserved by Kyro EOB. Most bands common to patterns K and D correspond to ribosomal proteins of bound poly-somes.

relationship to other cellular components. Although quantitative differences in the concentration of several enzymes present in rough and smooth microsomes have been reported (Stetten and Ghosh, 1971; Okao and Omura, 1972), analysis by SDS acrylamide electrophoresis had demonstrated extensive similarities between the polypeptide composition of rough microsomal membranes stripped of ribosomes and membranes of smooth ribosomes (Neville, 1971; Dehlinger and Schimke, 1971; Kreibich and Sabatini, 1974). Recently, high resolution gradient SDS acrylamide gels have revealed that at least three peptides bands ($\sim$MW 65,000, 60,000 and 35,000) are present in rough microsomal fractions and absent in smooth microsomes (Fig. 4).

By comparative analysis with other subcellular fractions, we found that the peptide band with a mobility corresponding to a protein of mass $\sim$30,000 daltons corresponds to urate oxidase present in peroxysomal cores which contaminate samples of rough microsomes. The other two peptide bands ($\sim$MW 67,000 and 65,000) were shown to be characteristic components of the rough endoplasmic reticulum membranes. These peptides were solubilized from rough membranes by deoxycholate (DOC) and are not present in samples of bound polysomes obtained from rough microsomes treated with this detergent (Fig. 4D). They were recovered, however, with a high yield in a sedimentable subfraction which was obtained from rough microsomes by treatment with high concentrations (2.5×10^{-2}M) of the non ionic detergent Kyro EOB. This subfraction was almost completely devoid of lipids, contained the bound ribosomes and was highly enriched for the two polypeptides (Fig. 4K). Profiles obtained by sedimentation analysis in sucrose density gradients (Fig. 5), suggested that the subfraction contained putative large polysomes to which the polypeptides were bound. Mild RNase treatment of the samples did not affect the sedimentation patterns demonstrating that they did not correspond to true polysomes. On the other hand, incubation in a medium containing high salt or mild digestion with proteases led to the breakdown of the apparent polysome structures into small polysomes. This suggests that in the original structure ribosomes were joined together not only by the polysomal mRNA but also by links of a protein nature.

Electron microscopic examination of the Kyro EOB residue (Fig. 6) showed that the putative polysome patterns were produced by groups of ribosomes attached to membrane remnants which have the filamentous appearance of a protein network. In many instances a trilaminar structure resembling that of the original membranes could still be observed with the distance between adjacent ribosomes attached to the vestigial rough membranes being considerably smaller than in the original rough microsomes.

The two polypeptides contained in the ribosome-bearing network behaved as integral membrane components since they were not removed by previous treatment of the microsomes with high salt or by incubation in alkaline (0.1 N NaOH) or acid (0.5 N acetic acid) solutions. Only one of these polypeptides has been found to be exposed on the cytoplasmic face of intact rough microsomal membranes where it was accessible to both proteases and to a lactoperoxidase catalyzed iodination. Preliminary double-labeling experiments, in which RM are first labeled with ^{131}I and reiodinated with ^{125}I after removal of the membrane-bound ribosomes, indicate that these two peptides

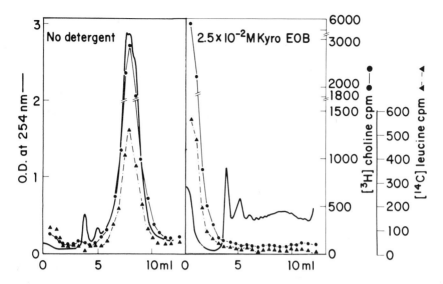

Figure 5. Sedimentation patterns obtained from RM and RM treated with Kyro EOB. RM containing phospholipids labeled *in vivo* ([³H] choline - 4 hr) and rapidly labeled proteins of the cisternal content ([¹⁴C] leucine - 30 min) were resuspended (3 mg protein/ml) in a low salt buffer. 450μl aliquots received 50μl each of water or Kyro EOB (2.5 x 10⁻¹M). Samples were layered onto linear sucrose density gradients (10-60% S-LSB). After centrifugation (90 min - 280,000 g - 4°) gradients were fractionated to determine TCA precipitable ¹⁴C - radioactivity in 100μl aliquots. ³H-radioactivity was measured directly in 100μl after solubilization in NCS. Sedimentation was from left to right.

In the left panel, intact microsomes containing labeled phospholipids and content proteins band isopycnically in the lower half of the gradients. After treatment with Kyro EOB, both phospholipids and content proteins are released and found on top of the gradient. Bound polysomes are displayed in a polysome-like pattern which is due to their association with a proteinaceous network derived from the membranes

are not part of the binding site which is covered by ribosomes in intact rough microsomes. They are, however, related to the binding sites and interconnect them forming an intramembranous scaffolding. One may speculate that the association of the binding sites with the network of proteins may be regulated intracellularly by factors which control the lateral mobility of the ribosomes. Moreover, the extent of the network would serve to limit the areas of membrane domains accessible to ribosomes.

Relationship of mRNA to ER Membranes

In a tissue culture system where it is possible to differentially label mRNA and ribosomes we investigated the possibility that mRNA of bound polysomes is directly associated with the ER membranes, independently of the large ribosomal subunits and the peptidyl tRNA.

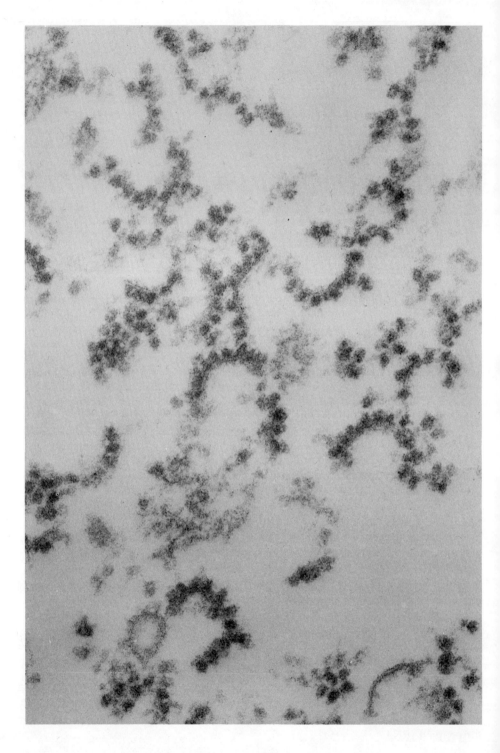

The WI-38 cell strain of normal human diploid fibroblasts which was selected for these studies secretes significant amounts of collagen into the culture medium (Houck, Sharma and Hayflick, 1971). When the cells have reached confluency, more than 40% of the cytoplasmic ribosomes (Tashiro *et al.*, 1975), are bound to a well-developed rough endoplasmic reticulum.   After cell disruption, a crude membrane fraction containing rough microsomes as well as smooth vesicles, plasma membrane fragments and mitochondria can be recovered by zonal centrifugation.   Membrane fractions sedimented through sucrose gradients containing high KC1 concentrations and Mg^{2+} ions are devoid of contaminating or adsorbed free ribosomes and polysomes and contain only active cytoplasmic ribosomes which are tightly bound to the microsomal membranes through their nascent polypeptide chains.

The fate of mRNA was examined after the *in vitro* disassembly of bound polysomes using membrane fractions obtained from cells which had been double-labeled by adding $[^{14}C]$ RNA precursors during growth to label ribosomal RNA, followed by ^{3}H-labeled precursors added in the presence of actinomycin D for a short period before harvesting the cells to label the mRNA. We found that, in spite of the extensive ribosomal release and discharge of nascent chains caused by treatment with either high salt media lacking Mg^{2+} (Fig. 8) or high salt-puromycin (Fig. 7), the labeled mRNA always remained attached to the sedimentable, ribosome-stripped membrane vesicles. Independent measurements of the distribution of poly (A)-containing mRNA purified by oligo(dT) chromatography after the disassembly of bound polysomes (Fig. 9) confirmed these results. Selective digestion of RNA under conditions in which poly (A) is not degraded showed that poly (A) segments of 150-200 nucleotides in length were retained within the membranes (Fig. 8, bottom panel) but were not present in the released material which consisted mainly of ribosomal subunits and some messenger fragments.   The poly (A) of membrane-associated mRNA remained bound to the membranes even after these membranes were incubated with pancreatic or T1 RNase under conditions which led to the complete digestion of the non-poly (A) portion of the messenger.   This membrane-associated poly (A) was easily digested, however, when intact or stripped membranes were incubated with pancreatic RNase in low ionic strength solutions or with T2 RNase suggesting that the 3' end of the bound messenger is exposed to the cytoplasmic aspect of the ER membranes.   In addition, mRNA molecules containing labeled poly (A) segments were easily removed from intact or stripped membrane vesicles by treatment with proteolytic enzymes; a finding which also suggests that proteins may be involved in this binding.

⟨——————

Figure 6.   Thin section electron micrograph of the sedimentable subfraction obtained from rough microsomes.   The subfractions were treated in a low salt buffer with 2.5 X 10^{-2}M of Kyro EOB. The sample was fixed for 30 min at 0° in 1% glutaraldehyde containing 2.5 X 10^{-2}M Kyro EOB, collected by filtration through a Millipore filter (Baudhuin *et al.*, 1967) and post fixed with 2% $0s0_4$ (227,500X).

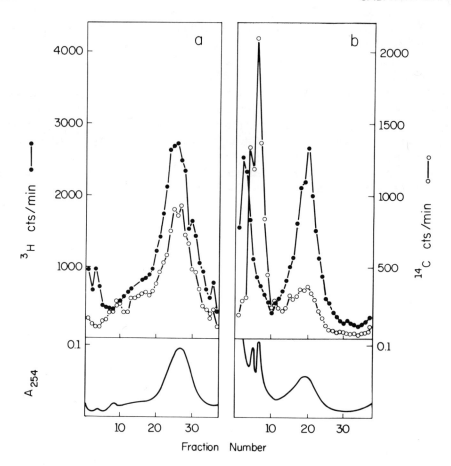

Figure 7. Selective release of ribosomal subunits by puromycin-KCl from a membrane fraction containing labeled messenger [³H] and ribosomal [¹⁴C] RNA. A culture was labeled with [¹⁴C]uridine (1 μCi/roller bottle; sp. activity 55 mCi/mmole) added 72 hr before confluency. After a medium change, actinomycin D was added to a final concentration of 0.03 ug/ml; 30 min later [³H]uridine (1 μCi/roller bottle, sp. activity, 26 Ci/mmole) was added and the culture was harvested after 3 hr. The cells were swollen in an hypotonic medium for 10 min and then broken with ten strokes of a tight fitting glass Dounce homogenizer. A postnuclear supernatant (PNS) was obtained by centrifugation (850 g - 2 min) and adjusted to the ionic composition of HSB (500 mM KCl, 50 mM Tris-HCl pH 7.4, 5 mM mgCl₂). A membrane fraction (MF) was obtained as a pellet by centrifugation of the PNS in a 15-30% sucrose density gradient containing HSB for 30 min at 25,000 rpm in a SW 41 Beckman rotor at 3°. The pellets, resuspended in HSB, were divided into two aliquots; one was treated with puromycin (10⁻³M) in HSB for 15 min at 37° for ribosomal stripping (b), and the other used as control (a). Both samples were analyzed by sedimentation at 20° (1 hr-40K-SW41) in 20-50% sucrose gradients containing HSB.

Optical density profiles were recorded (lower panels) and the radioactivity distributions were measured for each gradient. Sedimentation was from left to right.

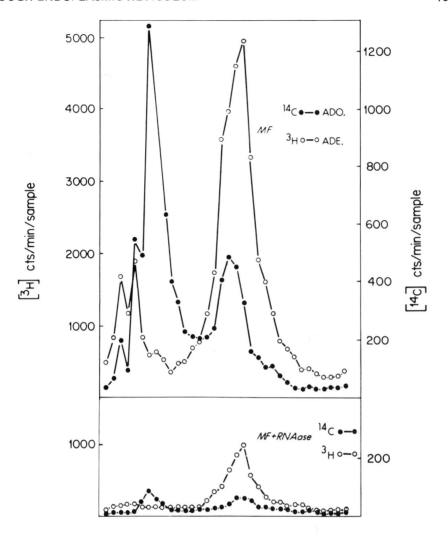

Figure 8. Retention of the RNase resistent poly A segments of mRNAs on membranes stripped of ribosomes by treatment with a high salt medium lacing Mg^{2+}. Cultures double-labeled in the ribosomal ([^{14}C] adenine, 1 μCi/roller bottle) and mRNAs ([^{3}H] adenosine, 1 mCi/roller bottle) as described in the legend to Fig. 7 were used to prepare a membrane fraction (see legend to Fig. 7) which was treated for ribosome stripping in a medium of high salt containing no Mg^{2+} (1.0 M KCl, 0.01 M Tris-HCl, pH 7.6) and then fractionated in 20-60% sucrose gradient of the same salt composition. Aliquots (100 μl each) of each fraction were diluted 10 times with water and treated with 0.5% sodium deoxycholate. One aliquot of each fraction was used to measure the ^{14}C- and ^{3}H-cold-acid-precipitable radioactivity (top panel). The other aliquot received pancreatic ribonuclease (2.5 ug/ml) and Tl ribonuclease (5 units/ml) and was incubated for 30 min at 30° before the measurement of acid-precipitable radioactivity (bottom panel). Sedimentation was from left to right.

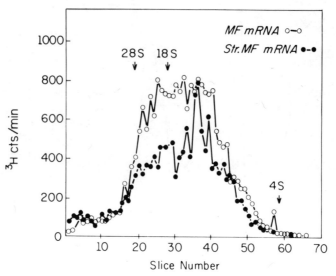

Figure 9.   Retention of poly (A)-containing mRNA on membranes stripped of ribosomes.   After medium change a confluent culture was labeled for 3 hrs with [^{3}H] adenosine (1 mCi/bottle, sp. activity 26 Ci/mmole) without addition of actinomycin D. Control (MF) and ribosome stripped (str. MF) fractions were prepared as indicated in the legend to Fig. 8.   RNA was extracted from the membrane fractions and poly (A)-containing molecules were purified by oligo-(dT) chromatography and analyzed by electrophoresis in a composite polyacrylamide (2.6%) agarose (0.5%) gel for 3 hr at 3 milliamperes per tube.   Gels were sliced and the distribution of the radioactivity determined.

The possibility that a significant fraction of the membrane-associated mRNA was of mitochondrial origin was ruled out by the following observations: (1) the accessibility of poly (A) to nucleases, (2) the size of the poly (A) segments associated with membranes was larger than the size of the poly (A) in mitochondrial RNA and (3) the labeling of the membrane associated mRNA was sensitive to campthothecin but not to ethidium bromide.   Mixing experiments in which messengers of added polysomes did not bind to the membranes during the ribosome stripping procedure rendered unlikely the possibility that the mRNA was artifactually adsorbed to the membranes after its release from the polysomes.   We therefore concluded that mRNA of bound polysomes contains at least one site for direct attachment to ER membranes and that this site is located near the poly (A) segment at the 3' end of the messenger.   A similar conclusion was recently reached by Milcarek and Penman (1974) using HeLa cells.   It is possible that in other systems the binding of mRNA to the membranes is weaker.   This may explain the finding (Harrison *et al.*, 1974b) that in myeloma cells the immunoglobulin mRNA is released from the membranes during polysome disassembly.

Our observations concerning the distribution of poly (A) after ribosome release and/or treatment with RNase should not be taken to indicate that the poly (A) segments by themselves serve to determine the binding of the messenger to the membranes since mRNA of both free and bound polysomes contain poly (A) segments of similar size at their 3' ends. The electrophoretic mobility of the poly (A) segments recovered from membranes incubated with RNase was found to be similar to the mobility of poly (A) segments prepared from RNA which was first extracted from the membranes, purified, and then digested with RNase. The resolution of polyacrylamide gel electrophoresis, however, is not sufficient to exclude the possibility that the poly (A) extracted from the membranes contains additional oligonucleotide sequences, which are protected from RNase during digestion of membrane vesicles and are responsible for the binding to the membranes. Such sequences may constitute direct recognition sites for the membranes or, more likely, for proteins which serve to bind the mRNA to the ER membranes. One may speculate that, in addition to proteins associated with poly (A) (Kwan and Brawerman, 1972; Blobel, 1973; Lindberg and Sundquist, 1974), other proteins specific for messengers in free and bound polysomes may be associated with regions adjacent to the poly (A) and perhaps with other regions of the mRNA which could contribute to determining its subcellular distribution.

In agreement with the conclusion that mRNA in bound polysomes is directly attached to ER membranes, other experiments have suggested the existence of a population of mRNA molecules in non-stripped membrane fractions which is bound to the membranes without being associated with ribosomes. A considerable fraction ($\sim$40%) of the mRNA labeled in the presence of actinomycin D was not recovered with the bound polysomes obtained after detergent treatment, but with the solubilized membrane material found in the top portions of the gradients. Although part of this "non-ribosome-associated" mRNA may represent degradation of polysomal mRNA this observation raises the possibility that even smooth portions of the endoplasmic reticulum, which do not contain bound ribosomes, may nevertheless contain some membrane-associated messenger RNA.

A Model for Translation in Bound Polysomes

A schematic model for the process of assembly of membrane-bound polysomes which incorporates the features described in previous sections and the binding of mRNA to the membranes through a segment near the 3' end is presented in Figure 10. Two possibilities are indicated in which initiation complexes formed near the free 5' end of the messenger bind to large subunits which may exist either free in the cytoplasm (Fig. 10a) or already bound to membranes at the specific receptor sites (Fig. 10b). In the first case, short nascent chains may play a role in the recognition of specific membrane binding sites for the active monomers. Only bound ribosomes which contain sufficiently long nascent peptides are tightly anchored to the specific membrane receptor sites by their extra-ribosomal amino terminal segments (Sabatini et al., 1972; Milstein et al., 1972; Brownlee et al., 1972; Harrison et al., 1974a, b). In the presence of Mg^{2+} at high ionic strengths in vitro, such ribosomes would remain associated with the membranes and would not be released by an RNase treatment which cleaves the mRNA. Other ribosomes within

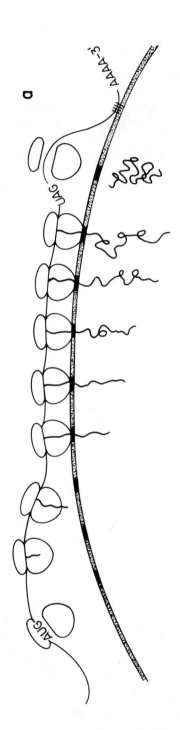

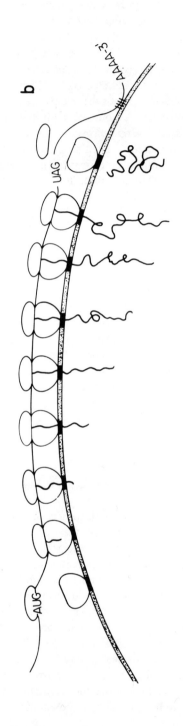

the same polysomes which are closer to the 5' end and contain short nascent chains may be more easily detached from their binding sites and, hence, released by mild RNase digestion.

A type of membrane-bound polysome has been described in cultured transformed cells in which numerous ribosomes are released from membranes by mild RNase treatment (Rosbash and Penman, 1971; Lee et al., 1971). Polysomes of this type which contain only or mainly "loose" or dangling ribosomes may be formed in unbalanced situations in which there are more ribosomes involved in the translation of membrane-associated messengers than there are available membrane sites for ribosome binding. One should note, however, that "loose" polysomes may also be formed artifactually when, during cell fractionation at low ionic strengths, free polysomes have an opportunity for adsorption onto available binding sites on membranes (Borgese et al., 1974). It is likely that polysomes described by Zauderer et al. (1973), which are released from membranes by washing in high salt media containing Mg^{2+} and synthesize the same proteins as free polysomes, result from this type of adsorption.

As discussed previously, a satisfactory model for protein synthesis in bound polysomes must contain features which allow for the movement of messenger RNA with respect to the ribosomes necessary for decoding. The possibility that extensive lateral displacement of binding sites for ribosomes and messenger is facilitated by the fluidity of the membrane and is controlled by a protein network in the membranes, was also discussed. Here we will consider an alternative mechanism (Figs. 11a, b) in which the messenger contains an untranslated region between the termination codon and the membrane messenger binding sites of sufficient length that, even if both the ribosome and the messenger binding sites are fixed in the membrane, the termination codon can still traverse all ribosomes within a polysome. In the extreme case in which the geometrical configuration of a polysome on a membrane is fixed, the length of the untranslated segment would determine the domain of the membrane to which a particular messenger RNA molecule is accessible for translation by bound ribosomes. It is important to note that the length of the untranslated segment would be minimized if the ribosomes are arranged in a circular "rosette" as shown in Figure 11, with the membrane-messenger binding site at the center of the rosette. Other polysome configurations leading to rosettes of different shapes would be possible only with longer untranslated messenger regions.

⟨————————

Figure 10. Scheme for assembly of bound polysomes with mRNA attached to the membrane through a segment near the 3' end. Ribosome binding to the membranes has been shown to occur via the large subunits and to be stabilized by sufficiently long nascent polypeptide chains which emerge from the subunits and penetrate into the membrane. Ribosomes near the 5' end, which contain short polypeptide chains can be found in vitro either dangling (a) or directly contacting the membrane (b), depending on the ionic strength.

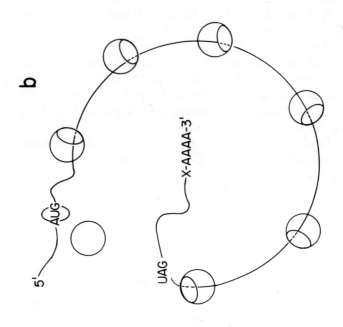

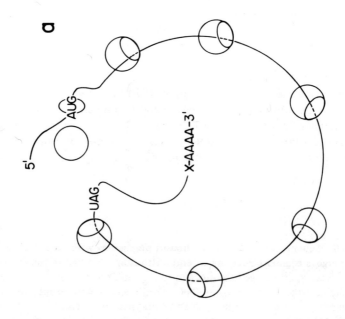

If, on the other hand, the length of the untranslated segment (r) is constant for all messenger lengths, then, as shown in Figure 12 for messengers with a translated region longer than $2\pi r$, a spiral-like arrangement, similar to the one commonly seen in electron micrographs, would be required to accommodate the additional ribosomes involved in translating the longer message.

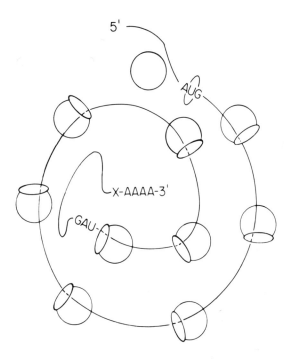

Figure 12. A spiral-like arrangement of bound ribosomes, similar to the one commonly seen in electron-micrographs of grazing sections of rough endoplasmic reticulum, would be required to accommodate the number of ribosomes which at any given time might be involved in translating a message 2π times longer than the untranslated segment at the 3' end.

⟨————————

Figure 11. Model explaining translation in bound polysomes with ribosomes as well as the 3' end of mRNA fixed on the ER membrane. In the schemes the membrane is represented by the plane of the page and the ribosomes are viewed from the top. An untranslated segment between the termination codon (UAG) and the messenger binding site to the membrane (X) allows termination and subsequent reinitiation in all ribosomes without detachment of the 3' end of the messenger from the membrane. a and b represent two successive steps in translation.

Some mechanism must insure that the initiation codon of a membrane-bound messenger is not utilized by ribosomes bound to the membrane beyond the domain allowed by the untranslated segment near the 3' end. Two possible mechanisms may accomplish this restriction. One is that there may be no ribosome-binding sites on the membrane which can be reached by the initiation codon (AUG) in the messenger at a distance from the anchoring site greater than the length of the constant region. Another may result from an attachment of the messenger region toward the 5' end of the initiation codon (proximal) to the region distal from the termination codon, which effectively would circularize the messenger and facilitate initiation in ribosomes which have just terminated.

Acknowledgements

This work was supported by grants GM 20277 and HD 06323 from the National Institutes of Health and by grant NP 129 from the American Cancer Society. The authors gratefully acknowledge the technical assistance of Miss Belinda Ulrich, Mrs. Marta Serpa and Mr. Bill Dolan and the photographic work of Mr. Miguel Nievas. Thanks are given to Mrs. Myrna Cort and Miss Susan Schwartz for the preparation of the manuscript.

REFERENCES

Adelman, M.R., Sabatini, D.D. and Blobel, G. (1973a). *J. Cell Biol. 56*, 206.

Adelman, M.R., Blobel, G. and Sabatini, D.D. (1973b). *J. Cell Biol. 56*, 191.

Akao, T. and Omura, T. (1972). *J. Biochem. 72*, 1245.

Baglioni, C., Bleiberg, I. and Zauderer, M. (1971). *Nature New Biology 232*, 8.

Baudhuin, P., Evrard, P. and Berthet, J. (1967). *J. Cell Biol. 32*, 18.

Blobel, G. (1973). *Proc. Nat. Acad. Sci. USA 70*, 924.

Blobel, G. and Sabatini, D.D. (1971a). *Biomembranes* vol. 2, ed. Manson, L.A. (New York: Plenum Pub. Co.), p. 193.

Blobel, G. and Sabatini, D.C. (1971b). *Proc. Nat. Acad. Sci. USA 68*, 390.

Borgese, N., Kreibich, G. and Sabatini, D.D. (1972). *J. Cell Biol. 55*, 24a.

Borgese, N., Blobel, G. and Sabatini, D.D. (1973). *J. Mol. Biol. 74*, 415.

Borgese, N., Mok, W., Kreibich, G. and Sabatini, D.D. (1974). *J. Mol. Biol. 88*, 559.

Branton, D. and Deamer, D.W. (1972). *Protoplasmatologia 11*, E1.

Brown, D.G. and Abrams, A. (1970). *Biochim. Biophys. Acta. 200*, 522.

Brownlee, G.G., Cartwright, E.M., Cowan, N.J., Jarvis, J.M. and Milstein, C. (1972). *Nature New Biology 244*, 236.

Burke, G.T. and Redman, C.M. (1973). *Biochim. Biophys. Acta. 299*, 312.

Chua, N., Blobel, G. and Siekevitz, P. (1973). *J. Cell Biol. 57*, 798.

Dehlinger, P.J. and Schimke, R.T. (1971). *J. Biol. Chem. 246*, 2574.

DeLorenzo, F., Goldberger, R.F., Steers, E., Givol, D. and Anfinsen, C.B. (1966). *J. Biol. Chem. 241*, 1562.

Diegelmenn, R.F., Bernstein, L. and Peterkofsky, B. (1973). *J. Biol. Chem. 248*, 6514.

Ekren, T., Shires, T. and Pitot, H.C. (1973). *Biochem. Biophys. Res. Comm. 54*, 283.

Florendo, N.Y. (1969). *J. Cell Biol. 41*, 335.

Fridlender, B.R. and Wettstein, F.O. (1970). *Biochem. Biophys. Res. Comm. 39*, 247.

Groner, Y., Pollack, Y., Berissi, H. and Revel, M.C. (1972). *Nature New Biol. 239*, 16.

Harrison, T.M., Brownlee, G.G. and Milstein, C. (1974a). *Eur. J. Biochem. 47*, 613.

Harrison, T.M., Brownlee, G.G. and Milstein, C. (1974a). *Eur. J. Biochem. 47*, 621.

Hochberg, A.A., Stratman, F.W., Sahlten, R.N., Morris, H.P. and Lardy, H.A. (1972). *Biochem. J. 130*, 19.

Houck, J.C., Sharma, V.K. and Hayflick, L. (1971). *Proc. Soc. Exp. Biol. Med. 137*, 331.

James, D.W., Rabin, B.R. and Williams, D.J. (1969). *Nature 224*, 371.

Jothy, S., Tay, S. and Simpkins, H. (1973). *Biochem. J. 132*, 637.

Karnovsky, M.J., Unanue, E.R. and Leventhal, M. (1972). *J. Exper. Med. 136*, 907.

Kellems, R.E. and Butow, R.A. (1972). *J. Biol. Chem. 247*, 8043.

Kellems, R.E., Allison, V.F. and Butow, R.A. (1974). *J. Biol. Chem. 249*, 3297.

Kellems, R.E. and Butow, R.A. (1974). *J. Biol. Chem. 249*, 3304.

Kuriyama, Y. and Luck, D.J. (1973). *J. Cell Biol. 59*, 776.

Kreibich, G. and Sabatini, D.D. (1973). *Fed. Proc. 32*, 9.

Kreibich, G., Debey, P. and Sabatini, D.D. (1973). *J. Cell Biol. 58*, 436.

Kreibich, G. and Sabatini, D.D. (1974a). *J. Cell Biol. 61,* 789.

Kreibich, G. and Sabatini, D.D. (1974b). *Methods in Enzymology 31,* 215.

Kreibich, G., Hubbard, A.L. and Sabatini, D.D. (1974). *J. Cell Biol. 60,* 616.

Kwan, S.W. and Brawerman, G. (1972). *Proc. Nat. Acad. Sci. USA 69,* 3247.

Lazarides, E.L., Luken, L.N. and Infante, A.A. (1971). *J. Mol. Biol. 58,* 831.

Lee, S.Y., Krsmanovic, V. and Brawerman, G. (1971). *J. Cell Biol. 49,* 683.

Lee-Huang, S. and Ochoa, S.C. (1972). *Biochem. Biophys. Res. Comm. 49,* 371.

Lindberg, V. and Sundquist, B. (1974). *J. Mol. Biol. 38,* 355.

Margulies, M.M. and Michaels, A. (1974). *J. Cell Biol. 60,* 65.

Milcarek, C. and Penman, S. (1974). *J. Mol. Biol. 89,* 327.

Milstein, C., Brownlee, G.G., Harrison, T.M. and Mathews, M.B. (1972). *Nature New Biol. 239,* 117.

Neville, D.M. Jr. (1971). *J. Biol. Chem. 246,* 6328.

Nolan, R.D. and Munro, H.N. (1972). *Biochim. Biophys. Acta. 272,* 473.

Nonomura, Y., Blobel, G. and Sabatini, D.D. (1971). *J. Mol. Biol. 60,* 303.

Nudel, N., Lebleu, B. and Revel, M. (1973). *Proc. Nat. Acad. Sci. USA 70,* 2139.

Ojakian, G., Kreibich, G., Kruppa, J., Mok, W. and Sabatini, D.D. (1975). In press—abstract of paper to be presented in April 1975 at Federation Meetings.

Palade, G.E. (1958) in *Microsomal Particles and Protein Synthesis,* ed. Roberts, Richard B. (New York: Pergamon Press), p. 36.

Palade, G.E. and Siekevitz, P. (1956). *J. Biophys. Biochem. Cytol. 2,* 171.

Pitot, H.C. and Shires, T.K. (1973). *Fed. Proc. 32,* 76.

Redman, C.M. and Cherian, M.G. (1972). *J. Cell Biol. 52,* 231.

Redman, C.M. and Sabatini, D.D. (1966). *Proc. Nat. Acad. Sci. USA 56,* 608.

Redman, C.M., Siekevitz, P. and Palade, G.E. (1966). *J. Biol. Chem. 241,* 1150.

Rolleston, F.S. (1972). *Biochem. J. 129,* 721.

Rolleston, F.S. and Lamm, T.Y. (1974). *Biochem. Biophys. Res. Comm. 59*, 467.

Rolleston, F.S. and Mak, D. (1973). *Biochem. J. 131*, 851.

Roobal, A. and Rabin, B.R. (1971). *FEBS Letters 14*, 165.

Rosbash, M. and Penman, S. (1971). *J. Mol. Biol. 59*, 227.

Sabatini, D.D. and Blobel, G. (1970). *J. Cell Biol. 45*, 146.

Sabatini, D.D., Blobel, G., Nonomura, Y. and Adelman, M.R. (1971) in *Advances in Cytopharmacology I: First International Symposium on Cell Biology and Cytopharmacology,* eds. Clemente, F. and Ceccarelli, B. (New York: Raven Press), p. 19.

Sabatini, D.D., Borgese, D., Adelman, M., Kreibich, G. and Blobel, G. (1972) in *RNA Viruses/Ribosomes.* North Holland FEBS Symposium *27,* 147.

Sabatini, D.D., Tashiro, Y. and Palade, G.E. (1966). *J. Mol. Biol. 19*, 503.

Schachter, H., Jabbal, I., Hudgin, R.L., Pinteric, L., McGuire, E.J. and Roseman, S. (1970). *J. Biol. Chem. 245*, 1090.

Scott-Burden, T. and Hawtrey, A.O. (1971). *Hoppe-Seyler's Z. Physiol. Chem. 352*, 575.

Scott-Burden, T. and Hawtrey, A.O. (1973). *Biochem. Biophys. Res. Comm. 54*, 1288.

Shelton, E. and Kuff, E.L. (1966). *J. Mol. Biol. 22*, 23.

Shires, T.K. and Pitot, H.C. (1973). *Nature New Biol. 242*, 198.

Shires, T.K. and Pitot, H.C. (1973). *Biochem. Biophys. Res. Comm. 50*, 344.

Shires, T.K., Narurkar, L.M. and Pitot, H.C. (1971a). *Biochem. J. 125*, 67.

Shires, T.K., Narurkar, L.M. and Pitot, H.C. (1971b). *Biochem. Biophys. Res. Comm. 45,* 1212.

Siekevitz, P. and Palade, G.E. (1960). *J. Biophys. Biochem. Cytol. 7*, 619.

Singer, S.J. and Nicolson, G.L. (1972). *Science 175*, 720.

Stetten, M.R. and Ghosh, S.B. (1971). *Biochim. Biophys. Acta. 223*, 163.

Sunshine, G.H., Williams, D.J. and Rabin, B.R. (1971). *Nature New Biol. 230*, 133.

Suss, R., Blobel, G. and Pitot, H.C. (1966). *Biochem. Biophys. Res. Comm. 23*, 299.

Szer, W. and Leffler, S. (1974). *Proc. Nat. Acad. Sci. USA 71*, 3611.

Tashiro, Y., Hadjiolov, A., Sumida, M., Lande, M. and Sabatini, D.D. In preparation.

Williams, D.J. and Rabin, B.R. (1969). *FEBS Letters 4,* 103.

Williams, D.J. and Rabin, B.R. (1971). *Nature 232,* 102.

Zauderer, M., Liberti, P. and Baglioni, C. (1973). *J. Mol. Biol. 79,* 577.

MEMBRANE FLUIDITY AND CELLULAR FUNCTIONS

S.J. Singer

Department of Biology
University of California at San Diego
La Jolla, California 92037

Introduction

An area of intense current interest in molecular and cell biology is the structure of biological membranes. A great change in our picture of membranes has occurred in the last few years, and there is now fairly widespread acceptance of a model for the organization of the lipids and proteins of membranes called the "fluid mosaic model" (Singer and Nicolson, 1972). In this model (Fig. 1) the proteins that are **integral** to the membrane (Singer, 1971) are proposed to be globular molecules which are partly embedded in the membrane lipid, and partly protrude from it. This partial embedding is determined thermodynamically by the **amphipathic** character of the integral protein molecule; its hydrophobic end is embedded in the hydrophobic membrane interior and its hydrophilic end protrudes into the aqueous phase. The lipid, arranged largely as a bilayer, forms the matrix of the membrane, and since at physiological temperatures the lipid of most functional membranes is largely fluid, the integral proteins are in principle free to move about laterally and rapidly in the plane of the membrane. There is now a very substantial body of evidence that is consistent with, and strongly supports, the fluid mosaic model. The discussion of this evidence could easily occupy my entire presentation, but I thought that in the context of this symposium it might be more useful to adopt the model as a working hypothesis of membrane structure and consider what it might imply about the mechanisms of a variety of important cellular functions and activities.

In particular, I would like to discuss what roles membrane fluidity might play in cellular phenomena and their control. I think it has come as a great surprise to most biologists that membranes often do have such fluidity, and that various molecules can move around in the membrane at such rapid rates (for some recent reviews see Gitler, 1972; Singer, 1974; Edidin, 1974). Biologists have much reason to appreciate the nearly ubiquitous existence of highly ordered and reproducibly formed structures, whether at the molecular level with individual proteins and DNA, or in the higher realm of organismal morphology.   It therefore seems almost to be a violation of some basic concepts of biological structure that such a vital organelle as the cell membrane is basically a two-dimensional solution with no long-range order to its structure. Such plasticity of structure must clearly be fundamental; it cannot be simply adventitious.   There must be ample reasons why the functions carried out by membranes require such fluidity and molecular mobility, but these reasons are not clearly established at the present time.   What I want to do, therefore, is to explore with you a few processes of great interest in cell biology whose mechanisms may depend critically on the fluidity of membranes.   The three specific processes I will discuss briefly are (1) cell activation, (2) protein transport across membranes, and (3) membrane fusion.   I will then consider the control and the regulation of molecular mobility in membranes, and its possible roles in cellular phenomena.

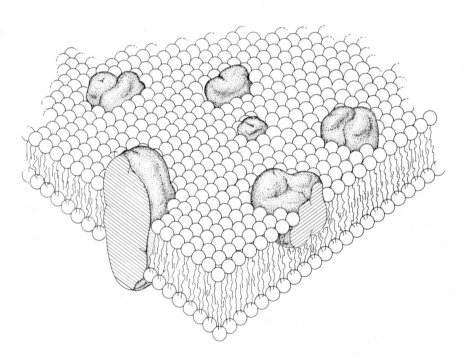

Figure 1.   A schematic representation of the fluid mosaic model of membrane structure (Singer and Nicolson, 1972).   The solid bodies with the stippled surfaces represent the integral protein molecules, which are distributed at random in the plane of the membrane within a lipid matrix.   If the lipid bilayer is in a fluid, or at least partly fluid state, the membrane structure is essentially that of a two-dimensional solution.

Cell Activation

There are a number of important systems where cells are activated by external agents to proliferate and differentiate. An antigen activating its specific unprimed lymphocyte, a mitogen or a peptide hormone activating its target cell, are just two examples. In these cases, the binding of the external agent to some receptor molecules in the exterior surface of the cell membrane initiates the activation process. It has been shown in several cases that the external agent does not have to get inside the cell in order for activation to occur. How does this binding lead to activation? One mechanism that has been widely entertained may be termed "trans-activation" (Singer and Nicolson, 1972). In this view the binding of the external agent to the receptor leads to some conformational change in the receptor which is somehow transmitted across the membrane in the localized region where the agent is bound. This mechanism may be the correct one in some cases, but there is not much evidence one way or the other at present. However, another possible mechanism has been suggested more recently by experiments on lymphocyte activation by antigens and mitogens. The plasma membrane of a B lymphocyte contains about 10^5 copies of an integral protein which is an immunoglobulin (Ig) molecule (Katz and Benacerraf, 1972). Each B lymphocyte has a different membrane-associated Ig. A particular antigen becomes bound to those lymphocytes whose Ig (acting as a receptor) has a specific affinity for that antigen. If the antigen is multivalent, that is, it has many identical sites per molecule, it can directly activate those B lymphocytes to which it is appropriately bound. (The antigen is called a T-cell independent antigen in these cases.) On the other hand, if the antigen has no, or only a few, identical sites, it cannot activate B cells directly, but requires the mediation of T cells in some more indirect and complex process.

Now, the evidence is that the specific binding of such multivalent antigens to specific B cells is generally accompanied by a rapidly-induced clustering of the receptor Ig (antibody) molecules in the plane of the cell membrane (Diener and Paetkau, 1972; Dunham et al., 1972). In effect, two-dimensional aggregation and "precipitation" of the multivalent antigen and its specific receptor Ig molecules is produced in the membrane. This clustering is not induced at 4^0, but occurs at 37^0, indicating that the fluidity of the membrane is required to allow adequate mobility of the receptor Ig molecules in the plane of the membrane.

Such observations suggest the possibility that the clustering induced upon antigen binding is somehow critical to triggering the lymphocyte activation process that follows. How a clustering of membrane receptors could lead to the next step in activation is not clear, but it may involve small but significant changes in the permeability of the membrane to specific ions or metabolites (such as Ca^{2+}) (Rasmussen et al., 1972) resulting in transient changes in the intracellular concentration of one or more critical ions or molecules which trigger the process. More generally, activation by antigens, mitogens, and hormones may involve such redistributions of components in the plane of the membrane (which we have termed "cis-activation") (Singer and Nicolson, 1972). A proposal of this sort has recently been made by Cuatrecasas for the mechanism of activation of cells by hormones (Cuatrecasas, 1974), and may turn out to be generally true of a wide range of activation processes.

An interesting system in which these mechanisms may be more accessible experimentally is the antigen-induced release of histamine from mast cells (for review see Becker and Hensen, 1973).

Protein Transport Across Membranes

After the binding of a multivalent antigen to its receptor Ig on a B lymphcyte surface membrane, or of an anti-Ig antibody to the receptor Ig, the redistribution that results often sweeps all of the receptor Ig molecules into one area of the cell surface ("capping"). At 37°, this may occur within a few minutes. Following upon this capping of the receptors, the capped regions of the plasma membrane are endocytosed in an energy-dependent process, forming intracellular vesicles. It may take a half hour or so at 37° to. sweep essentially all of the receptor out of the plasma membrane by this process. What molecular events are involved in the endocytotic step are not known, but some kind of actomyosin-like contractile protein system attached to the cytoplasmic surface of the membrane may be involved.

The evidence suggests that capping and its subsequent endocytosis are not essential to cell activation; on the contrary, they seem to inhibit it (Singer, 1974). If a redistribution of receptors in the membrane is indeed crucial to cell activation, therefore, the redistribution must be less extensive than is exhibited in capping. However, the capping and endocytosis that follows it are still very interesting phenomena and probably reflect mechanisms that are physiologically important. Such mechanisms may be operating, for example, in at least certain instances of the transport of specific proteins across cell membranes.

It is entirely unlikely thermodynamically that soluble proteins can simply penetrate cell membranes that are impermeable to much smaller hydrophilic molecules, and some specific mechanism must operate to facilitate such transport. I have suggested (Singer, 1974) that the general mechanism involved is very like that of lymphocyte capping and endocytosis (Fig. 2). The particular membrane must have a receptor protein in it capable of binding the specific protein to be transported (the ligand). It is postulated that the binding of the ligand to the receptor is then followed spontaneously by the redistribution and clustering of the bound receptors into local regions of the membrane. The clustering results either from the fact that the ligand molecule has multiple sites for binding to the receptor, or from an altered interaction of receptors in the membrane (Singer, 1974). The clustering then triggers the endocytosis of the clustered regions of the membrane, in an energy-dependent process that may involve an actomyosin-like system associated with the membrane.

There are two very different cases reported in the literature which appear to fit at

ABSORPTION OF PROTEINS ACROSS MEMBRANES

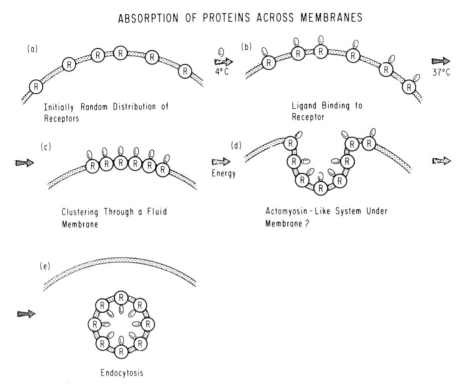

Figure 2. A generalized hypothesis for the mechanism of uptake of specific proteins (L) by cells (see Singer, 1974). The plasma membrane of the cell is presumed to contain receptor molecules (R) which are integral proteins specifically capable of binding L at the exterior face of the membrane. R molecules may or may not span the thickness of the membrane. The binding of L to R (panel a) leads to a clustering of bound receptors at 37° (panel c). Following this an energy-dependent process leads to the formation of a pinocytotic vesicle which thereby transports L molecules inside the cell. This energy requiring process may involve an actomyosin-like system attached to the cytoplasmic surface of the plasma membrane. This scheme is closely analogous to the "capping" and endocytosis that occur upon binding of a multivalent antigen to specific unprimed B lymphocytes (Diener and Paetkau, 1972; Dunham et al., 1972); it may also be closely related to the mechanism of phagocytosis (Singer, 1974).

least some of the expectations of this proposed mechanism. One case involves the uptake of ferritin by pro-erythroblasts. Considerably before the idea of a fluid membrane became current, Fawcett (1965) had shown that ferritin bound to the surface of a pro-erythroblast is collected into clusters at invaginations of the surface of the cell as well as in intracellular vesicles. Among the predictions we make is that if the same experiment were carried out at 4°, the ferritin would be found uniformly distributed over the surface, but would become clustered and endocytosed only upon raising the temperature to 37°, in analogy to the lymphocyte capping experiments. Another case involves the uptake of maternal antibodies through the intestine of the neonatal rat (Rodewald, 1973). The specificity of this process is quite striking. First of all, it is only the proximal one third of the neonatal intestine where this uptake occurs, and it no longer occurs there ten days after birth. Secondly, the species of immunoglobulin molecules is highly restricted. IgG of the rat, but not of the chicken, will be transported. This information suggests that a specific receptor for the rat IgG molecule must be present in the particular cell membrane involved, which turns out to be the brush border of the epithelial cells lining the intestine. Rodewald (1973) has shown that at the bases of the microvilli of the brush border the rat IgG molecules are specifically bound and clustered, and below the microvilli, endocytotic vesicles containing the clustered IgG are found.

These are just two cases where evidence has been obtained for what I think is likely to be a very wide-spread mechanism for taking up specific proteins into cells or organelles which require these externally-produced proteins for their subsequent development. Perhaps a mechanism such as this operates to transfer the soluble enzymes found in inner matrix of the mitochondrion from their sites of synthesis in the cytoplasm, across the inner mitochondrial membrane, although I know of no evidence for this at present. This same clustering and endocytotic mechanism may also operate in phagocytosis by macrophages and other phagocytic cells (Singer, 1974). If so, then the fluidity of the membrane and the lateral mobility of its receptors, are obligatory for such phenomena to occur.

Membrane Fusion

The mechanisms that operate in physiologically important membrane fusions (such as occur in the penetration of animal cells by viruses, and in a whole host of secretory processes involving the extrusion of secretory granules) are not known. One view of membrane fusion (see *e.g.* Lucy, 1970) is that it is essentially a spontaneous process that could occur between any two phospholipid-bilayer membranes unless there is some inhibitory factor (steric, electrostatic or other) preventing it. The specificity with which particular membranes fuse, in this view, is largely attributable to the absence of specific inhibitory factors. On the other hand, in one recent striking investigation of a membrane fusion process, a quite different general view of the fusion process is suggested. Satir *et al.* (1973) have studied the process of mucus secretion in *Tetrahymena,* which involves the fusion of intracellular vesicles called mucocysts with the plasma membrane. Their freeze-fracture electron micrographs are spectacular and their paper should be carefully studied. Their observations have shown that the sites of fusion on the plasma membrane are not random, but are specific regions characterized by a circular string of intramembranous particles (a "rosette"). The mucocyst membrane fuses with the plasma membrane only at these sites. At a stage in the overall process when the mucocyst is still quite a distance from the plasma membrane, the mucocyst membrane exhibits a random uniform distribution of its own intramembranous particles. But as the mucocyst comes to within a few hundred Angstroms of the plasma membrane, its intramembranous particles become arranged into an annulus which is spatially complementary to the rosette of intramembranous particles in the plasma membrane immediately adjacent. The fusion process almost certainly involves some interaction and union of the rosette and annulus particles into a single fused membrane. The molecular mechanisms responsible for these events are not known, but clearly involve the fluidity and molecular mobility of the membranes involved. How general this mechanism is to fusion is not clear. The geometry and frequency of fusion in the case studied by Satir *et al.* (1973) are very favorable; now that one knows what to look for, similar studies may be successful in less favorable membrane fusion systems.

The Control of the Mobility of Membrane Components

If the fluidity of membranes, and the molecular mobility of components in the plane of the membrane are important in the biology of cells, then the control of that fluidity and molecular mobility may also play a critical role. We know of many instances where there are organized structures within membranes, such as tight junctions, gap junctions, ciliary "necklaces" (Gilula and Satir, 1972), where it is clear that random diffusion of at least some of the membrane components is in some way restricted. How do such structures form in the membrane, and how are they maintained? We can only guess at the present time. The membrane of the adult human intact erythrocyte, remarkably enough, is a membrane whose components do not appear to be mobile. This is indicated by the fact that antibodies to surface antigens of the erythrocyte do not produce any apparent clustering, capping, or endocytosis of the antigen of the kind discussed above for lymphocytes and other cells (Blanton *et al.*, 1968; Loor *et al.*, 1972). Other kinds of experiments (see for example Peters *et al.*, 1974; Elgsaeter and Branton, 1974) suggest the same thing. On the other hand, with ghosts prepared from adult human erythrocytes, some mobility of the membrane components can be demonstrated (Pinto da Silva, 1972; Elgsaeter and Branton, 1974). This suggests that the restriction on mobility in the membrane of the intact cell is not due to a highly viscous lipid matrix, since in ghosting the lipid viscosity *per se* is not likely to change significantly, and, in fact, does not (Landsberger *et al.*, 1972). Furthermore, with newborn human intact erythrocytes, antibody-induced clustering and endocytosis has been demonstrated under conditions where no effects are produced with adult cells (Blanton *et al.*, 1968). All of these results suggest that the mobility of components in the erythrocyte membrane is under some control and can be regulated.

In our original formulation of the fluid mosaic model (Singer and Nicolson, 1972) we in fact suggested that in certain cases where molecular mobility in membranes was restricted, "some agent extrinsic to the membrane (either inside or outside the cell) interacts multiply with specific integral proteins," and thereby restricts the mobility of those proteins. What agent or agents might be involved in the erythrocyte membrane? It turns out that this membrane contains a large amount of a complex of two proteins called spectrin [bands 1 and 2 resolved in sodium dodecyl sulfate-polyacrylamide gel electrophoresis (Fairbanks *et al.*, 1971)] and another protein (band 5), which is among those proteins that are relatively weakly bound to the membrane ("peripheral" proteins) (Singer, 1971). We have obtained good evidence (Sheetz, Painter and Singer, to be published) that spectrin is a member of the family of muscle proteins, the myosins, while the protein of band 5 closely resembles the muscle protein actin. For example, rabbit antibodies to smooth muscle myosin isolated from the human uterus show a weak but specific cross reaction with human erythrocyte spectrin; while band 5 protein, when isolated and polymerized into fibers, forms arrowhead complexes with striated muscle heavy meromyosin, a very specific test for an actin-like molecule (Huxley, 1963). The erythrocyte membrane therefore has associated with it copious amounts of an actomyosin-like protein system.

How is this system attached to the membrane? Spectrin has been shown (Nicolson *et al.*, 1971) to be localized to the cytoplasmic surface of the membrane. Furthermore, there are experiments (Nicolson and Painter, 1973) which suggest that spectrin molecules are in some manner associated with the intramembranous particles in the membrane, presumably at sites where those particles protrude from the cytoplasmic surface of the membrane. These intramembranous particles span the membrane, and contain protein-bound oligosaccharide chains exposed at the exterior surface of the membrane (Marchesi *et al.*, 1972).

Our working hypothesis at present is that this actomyosin-like system of the erythrocyte membrane, while bound to the membrane, can undergo a reversible aggregation whose equilibrium state depends on the concentrations of certain soluble cytoplasmic components (perhaps Ca^{2+}, ATP, etc.). We have noted (Painter *et al.*, 1975) that such an aggregation-disaggregation process could provide a mechanism for the control of molecular mobility in the membrane. This concept is illustrated in highly schematic form in Figure 3. In the top panel of this figure, a peripheral component in its bound but disaggregated state is shown. In this state it has little or no influence on the distribution or mobility of the intramembranous particles to which it is attached. In its aggregated state (bottom panel), however, it effectively ties together a large cluster of intramembranous particles, inhibiting their mobility and perhaps that of other membrane components as well. I emphasize that this is a highly schematic picture, and the true situation might be more complex and involve several peripheral components as in an actomyosin-like complex, but the basic molecular mechanism that is proposed should be clear.

$\longrightarrow$

Figure 3. A highly schematic representation showing how a peripheral protein (or proteins) specifically bound to integral proteins at the cytoplasmic face of a membrane, in going from (a) its disaggregated state to (b) its aggregated state, could inhibit the translational mobility of the integral protein in the plane of the membrane.

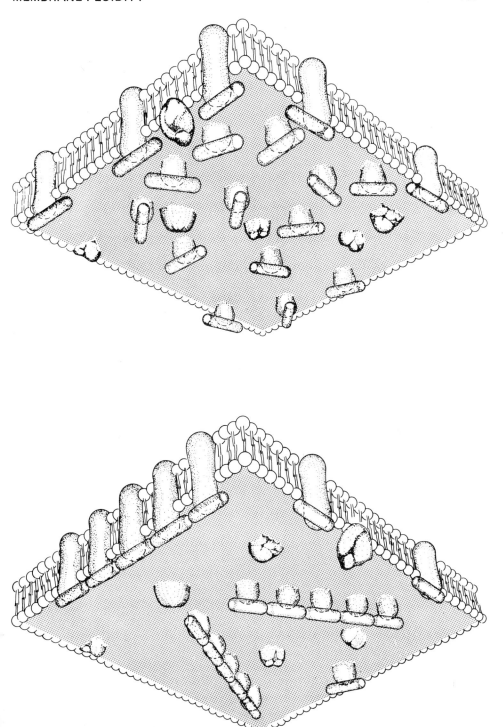

Among the many instances in cell biology where the control of molecular mobility in membranes may be important is in malignant transformation. Several years ago, considerable excitement was generated by the findings (Burger, 1969; Inbar and Sachs, 1969) that a variety of plant lectins, which are proteins capable of binding to specific sugar residues on glycoproteins and glycolipids, could cause malignantly transformed cells to agglutinate, but not their normal counterpart cells. Furthermore, if normal cells were treated mildly with trypsin, they now became readily agglutinable by the lectins. The explanation given for these results was that the plasma membrane of the normal cell had the lectin binding sites in a "cryptic" state, which were exposed upon trypsinization and upon malignant transformation. Thinking in terms of fluid mosaic model, however, it occurred to me that a different explanation would be tenable. Referring in particular to the difference in agglutinability of normal and trypsinized normal cells, I suggested that the numbers and crypticity of lectin binding sites might not be different, but that instead the sites might be clustered into patches in the membrane of the trypsinized cell and be molecularly dispersed in the membrane of the untreated cell. The probability of forming lectin bridges between such clusters of sites on two trypsinized cells would then be much greater than between isolated sites on two normal cells. By analogy, this picture was extended to the case of the malignantly transformed cell. These proposals were included in the original paper on the fluid mosaic model (Singer and Nicolson, 1972). A number of investigators (Nicolson, 1973; Rosenblith et al., 1973; Noonan and Burger, 1973) have since obtained results which confirm the essential correctness of this suggestion, but have shown that the clustering of lectin binding sites on malignantly transformed cells is **induced** by the binding of the lectin under physiological conditions, whereas no such clustering is induced in the normal cell. In other words, it appears that there is a difference in **mobility** of the lectin receptors in the two membranes: restricted in the normal cell membrane but relatively free in the malignantly transformed cell. Now, several explanations have been offered for this apparent difference in mobility. Among them are a difference in lipid composition and viscosity (Shinitzky and Inbar, 1974); and a difference in a surface protein component (Hynes, 1973; Stone et al., 1974; Hogg, 1974; Wickus et al., 1974; Gahmberg et al., 1974; Yamada and Weston, 1974) in the two membranes. Without discussing the pros and cons of these suggestions, I suggest still another possibility; namely that the difference in mobility is related to the scheme advanced in Figure 3; that there is a peripherally-bound, actomyosin-like protein system in these cells which undergoes reversible aggregation; that in the normal cell, its steady state is the highly aggregated one, and in the trypsinized normal cell and the malignantly transformed cell its steady state is a relatively disaggregated one. At the present time, we are trying to obtain experimental evidence for or against this hypothesis.

This suggested mechanism for the difference in surface membrane properties between normal cells and their malignantly transformed counterparts may turn out to have even more general import. It is conceivable that the control of the mobility of components in a membrane is the critical factor in the control of normal cell growth. A growing cell may require its plasma membrane components to be freely mobile so as to allow (1) the plasma membrane to grow by fusion of some precursor intracellular membranes with pre-existing plasma membrane (Palade, 1959); (2) an equal distribution of important plasma membrane components between two halves of a dividing cell; and (3) the process of cytokinesis to take place, among other special processes involved in cell growth. To stop growth, the crucial step may therefore be to immobilize membrane components by a mechanism analogous to that represented in Figure 3, *i.e.,* some signal may trigger the aggregation of an actomyosin-like protein system which is peripherally attached to the membrane. In this scheme, the malignantly-transformed cell is one which has lost the capacity to generate or respond to that signal, so that the actomyosin-like system associated with the membrane remains in its disaggregated state, the membrane components remain mobile, and the cell continues to grow.

REFERENCES

Becker, E.L. and Henson, P.M. (1973). *Advan. Immunol. 17,* 93.

Blanton, P.L., Martin, J. and Haberman, S. (1968). *J. Cell Biol. 37,* 716.

Burger, M.M. (1969). *Proc. Nat. Acad. Sci. USA 62,* 994.

Cuatrecasas, P. (1974). *Annu. Rev. Biochem. 43,* 169.

Diener, E. and Paetkau, V.H. (1972). *Proc. Nat. Acad. Sci. USA 69,* 2364.

Dunham, E.K., Unanue, E.R. and Benacerraf, B. (1972). *J. Exp. Med. 136,* 403.

Edidin, M. (1974). *Annu. Rev. Biophys. Bioeng. 3,* 179.

Elgsaeter, A. and Branton, D. (1974). *J. Cell Biol. 63,* 1018.

Fairbanks, G., Steck, T.L. and Wallach, D.F.H. (1971). *Biochemistry 10,* 2606.

Fawcett, D.W. (1965). *J. Histochem. Cytochem. 13,* 75.

Gahmberg, C.G., Kiehn, D. and Hakamori, S. (1974). *Nature 248,* 413.

Gilula, N.B. and Satir, P. (1972). *J. Cell Biol. 53,* 494.

Gitler, C. (1972). *Annu. Rev. Biophys. Bioeng. 1,* 51.

Hogg, N.M. (1974). *Proc. Nat. Acad. Sci. USA 71,* 488.

Huxley, H.E. (1963). *J. Mol. Biol. 7,* 281.

Hynes, R.O. (1973). *Proc. Nat. Acad. Sci. USA 70,* 3170.

Inbar, M. and Sachs, L. (1969). *Proc. Nat. Acad. Sci. USA 63,* 1418.

Katz, D. and Benacerraf, B. (1972). *Advan. Immunol. 15,* 1.

Landsberger, F.R., Paxton, J. and Lenard, J. (1972). *Biochim. Biophys. Acta 266,* 1.

Loor, F., Forni, L. and Pernis, B. (1972). *Eur. J. Immunol. 2,* 203.

Lucy, J.A. (1970). *Nature 227,* 815.

Marchesi, V.T., Tillack, T.W., Jackson, R.L., Segrest, J.P. and Scott, R.E. (1972). *Proc. Nat. Acad. Sci. USA 69,* 1445.

Nicolson, G.L. (1973). *Nature New Biol. 243,* 218.

Nicolson, G.L., Marchesi, V.T. and Singer, S.J. (1971). *J. Cell Biol. 51,* 265.

Nicolson, G.L. and Painter, R.G. (1973). *J. Cell Biol. 59,* 395.

Noonan, K.D. and Burger, M.M. (1973). *J. Cell Biol. 59,* 134.

Painter, R.G., Sheetz, M. and Singer, S.J. (1975). *Proc. Nat. Acad. Sci. USA 72,* 1359.

Palade, G.E. (1959) in *Subcellular Particles,* ed. Hayashi, T. (New York: Ronald Press), p. 64.

Peters, R., Peters, J., Tews, K.H. and Bahr, W. (1974). *Biochim. Biophys. Acta 367,* 282.

Pinto da Silva, P. (1972). *J. Cell Biol. 53,* 777.

Rasmussen, H., Goodman, D.B.P. and Tanenhouse, A. (1972). *C. R. C. Critical Rev. Biochem. 1,* 95.

Rodewald, R. (1973). *J. Cell Biol. 58,* 189.

Rosenblith, J.Z., Ukena, T.E., Yin, H.H., Berlin, R.D. and Karnovsky, M.J. (1973). *Proc. Nat. Acad. Sci. USA 70,* 1625.

Satir, B., Schooley, C. and Satir, P. (1973). *J. Cell Biol. 56,* 153.

Shinitzky, M. and Inbar, M. (1974). *J. Mol. Biol. 85,* 603.

Singer, S.J. (1971) in *Structure and Function of Biological Membranes,* ed. Rothfield, L.I. (New York: Academic Press), p. 145.

Singer, S.J. (1974). *Advan. Immunol. 19,* 1.

Singer, S.J. and Nicolson, G.L. (1972). *Science 175,* 720.

Stone, K.R., Smith, R.E. and Joklik, W.K. (1974). *Virology 58,* 86.

Wickus, G., Branton, P. and Robbins, P.W. (1974) in *Cold Spring Harbor Conference on Control of Proliferation in Animal Cells,* eds. Clarkson, B. and Beserga, R., p. 541.

Yamada, K.M. and Weston, J.A. (1974). *Proc. Nat. Acad. Sci. USA 71,* 3492.

BIOCHEMICAL FUNCTION AND HOMEOSTASIS:
THE PAYOFF OF THE GENETIC PROGRAM

Daniel E. Atkinson

Biochemistry Division
Department of Chemistry
University of California
Los Angeles, California 90024

Introduction

Design of living organisms by mutation and selection is in principle very similar to engineering design. In both cases changes are evaluated and those that are advantageous serve as the new basis for further testing of additional changes; in both cases improved function is a criterion on the basis of which changes are accepted or rejected. Thus when we deal with objects of biological origin or with objects resulting from human design our approach must be intellectually similar. It must be totally different from the approach that is appropriate when we deal with rocks, continents, or other objects of non-biological, non-designed systems. A functional object must have a design. The design need not necessarily be recorded: you could set out to build a bird house and begin with the floor and then make the sides and roof to fit. A generalized design in that case would be in your mind. However, when an object is complex or when many copies of the object are to be made, some means of recording the design is desirable. Thus we have blueprints and the like, which may themselves become extensive and elaborate. I remember reading somewhere that by the time a battleship was built the paper used for plans or blueprints of the ship and its component parts weighed about as much as the ship itself. There is probably some exaggeration in that statement, but it may not be far wrong. At any rate, the battleship would not have been able to carry a really large number of total copies of its design without foundering. It is interesting that we constantly carry around several billion copies of the total design for our bodies. If we could write our genetic information on one sheet of typing paper, one billion copies would weigh about 5,000 tons. Genetic programs are indeed pretty well miniaturized.

The point that I want to emphasize about this miniaturized evolutionary program, or any other design, is that the designed object is the object of the design. That is not just a play on words; blueprints or any other plans are means to an end; they are not ends in themselves. You are much more interested in your color television or your Volkswagen than you are in the drawing that leads to the television or the automobile. Because of the extreme interest in the genetic program and in the ways in which it is reproduced we sometimes, in biology, tend to invert this relationship. In legal jargon we might be said to stand the argument on its head. As a graduate student I saw a statement, which I must paraphrase because I have not been able to find it again, to the effect that chromosomes are the fragile canoes of life, voyaging down through the eons on a turbulent sea of cytoplasm. That's a nice poetic concept, and it obviously impressed me a great deal, since I remember it thirty years later. One must grant, however, that it expresses a certain parochialism as to what biology is about. The concept of germ cell line versus somatic cell line, when overemphasized, is similarly ambiguous and conceptually misleading. The success of the journey down through the eons depends on how well the cytoplasm of the somatic cells deals with the environment and with constantly changing threats and opportunities that are offered by the environment. The organism, in other words, is the organism. The organism is not the chromosome, any more than the automobile is the blueprint.

Homeostasis

In my title I used the term "payoff of the genetic system." The ultimate payoff is of course survival, reproduction, and having great-grandchildren. But at the operational or functional level I think it is reasonable to imply, as I did in the title, that biochemical homeostasis is the immediate payoff of the genetic program. If I were to pick one unique characteristic of life it would be homeostasis or regulation. Almost nothing else—catalysis, the ability to do work, etc.—is really unique. It is designed homeostasis that differentiates the living world (and some things designed by living organisms) from the inanimate world. The concept of homeostasis was developed by physiologists long ago, and was implicitly recognized much earlier. Regulation of body temperature in mammals and birds was of course familiar to prehistoric man, and loss of body heat is a sign of death.

Physiologists of the last century began to realize that the body fluids were closely regulated in terms of pH and ionic strength. But the basic homeostasis obviously must be that of the metabolites in the cytoplasm of the functioning somatic cell. A generalization that has emerged in the last few years, and one that I believe to be as fundamental as any concept in biology, is that the genetic program provides that metabolic fluxes will vary widely in order to maintain concentrations of metabolites within narrow limits. That is, cellular homeostasis is homeostasis of concentrations but of course not of fluxes.

Negative Feedback

One of the simplest types of regulatory systems, negative feedback by the end product of a biosynthetic sequence, was discovered in 1956 by Umbarger and by Pardee and is now familiar to all biologists. This is identical in principle to a thermostat or any other device in which fluctuations in some such parameter as temperature, pressure, or concentration are sensed and used to stabilize the value of that parameter. The concentration of an end product is sensed by the enzyme catalyzing the first committed step—the first reaction that is specific for making that product—and a consequence of this sensing is that when the product concentration is higher the enzyme will have a lower affinity for its substrate. Since this enzyme catalyzes the first committed step, its substrate is necessarily a branchpoint metabolite that is used also in at least one other sequence. Thus changes in affinity will lead to changes in partitioning of the metabolite between the alternative pathways. An important design feature of this kind of control is that the response is not like turning a water faucet on and off. The effect is much more like that resulting from vertical movement of a dam regulating flow into a lateral ditch of an irrigation system. The enzyme is not turned off; its affinity for the substrate is changed. This changes the effectiveness with which the enzyme competes with another enzyme for the common substrate. It is easy to show either graphically or algebraically that variation of affinities allows for much sharper changes in partitioning between metabolic pathways than could be obtained by changing the maximum velocity, or catalytic constant, of the enzyme.

In a sense a more fundamental, or at least a broader, question is how the rates of major sequences are correlated. Although they are found throughout biosynthetic metabolism, branch points at which partitioning is regulated by the concentration of end product could not, even in principle, provide all of the necessary control. End product negative feedback provides for response to needs for individual products, but cannot correlate needs with resources. A more general input is clearly necessary.

Adenine Nucleotides

It has been recognized for about thirty years that adenosine triphosphate, ATP, links metabolic reactions that may be loosely said to provide energy with those that require it. We have come to realize that the adenine nucleotide system provides a stoichiometric linkage between all biochemical sequences. This is the only metabolic system that is ubiquitous; that is, it is involved in all metabolic sequences. This necessarily requires that if the sequences are to be kinetically regulated and correlated, the control must be exerted through ATP. Such correlation cannot be obtained by response to any other system, unless that system is a secondary indication of the situation in the adenylate pool.

I have been using terms implying function and purpose in metabolism. This may be an appropriate time for comment on such usage. It is of course evident that the *facts* and *concepts* of chemistry and physics are essential to the study of living systems. But it may be less generally recognized that the traditional *approach* of chemistry and physics is fatal to the understanding of biological systems. Chemistry and physics have developed with primary concern for inanimate systems, and they rightly exclude "why" questions and ideas of function or purpose. These concepts are inapplicable when you study a hydrogen nucleus or a mineral crystal. They are, however, indispensable in the study of any living organism or its components. We can no more predict a biological system by simple extrapolation from the properties of the compounds of which it is composed than we could predict television by knowing some metallurgy and a little physics. Obviously the fundamental properties of matter are essential to a design but they do not make the design. We need the approaches of engineering and economics, or at least approaches that correspond intellectually to those of engineering and economics, in order to use the facts of chemistry and physics as aids to understanding biological systems.

Intelligent visitors from an advanced civilization on another planet might deduce a great deal about the history of the earth from a collection of earth rocks, just as human scientists have found the basis for much deduction and speculation in a small collection of rocks from the moon. But if these visitors happened to pick up an old spark plug, the femur of a fox, a typewriter ribbon, or a chiton shell, they could not, no matter how intelligent, understand much about these objects. Anything that is designed, whether by evolution or by a man or other organism, is understandable only in context, and a component of a complex system is understandable only in terms of its role in the intact system.

The standard multicolored wall charts of metabolic pathways illustrate convincingly that metabolism is complex, but they are not very suitable for showing functional relationships. For metabolizing cells, as for electronic devices, a simplified functional block diagram is a useful supplement to a detailed schematic diagram. Figure 1 is such a functional block diagram for a typical aerobic heterotrophic cell. The activities of such a cell may be grouped into three functional categories, as illustrated by the three blocks in the figure. Catabolism includes the sequences in which carbohydrates, fats, and proteins are oxidized to carbon dioxide. These sequences regenerate ATP, supply electrons to the NADP pool, and produce a surprisingly small number of compounds—nine or ten—that are the starting materials for all of the hundreds of biosynthetic sequences in the cell. The biosynthetic block represents many parallel pathways, of which the quantitatively most important are the synthesis of the amino acids and nucleotides, the building blocks for making macromolecules. The third block represents the production of macromolecules, assembly of supramolecular units and organelles, and the replication of cells. At a low level of resolution, this diagram represents the whole biochemistry of an aerobic heterotrophic cell. It is relevant here in particular because it shows the specific and unique role played by the adenine nucleotides and NADP as coupling agents. The biosynthetic block contains several hundred parallel pathways, each of which uses ATP. In many cases AMP instead of ADP is a product, but in the interests of simplicity the diagram shows only ADP. Most biosynthetic sequences contain reductive steps that use NADPH. ATP is regenerated from ADP or AMP through metabolic reactions, primarily in the central degradative pathways of glycolysis and the citrate cycle. NADPH is regenerated by reduction of $NADP^+$, at the expense of oxidation of substrate. NAD plays a similar coupling role in being reduced at the expense of substrate oxidation and then re-oxidized in connection with ATP regeneration, so that the same kind of regulatory interactions that we will discuss may be expected also to apply to the NAD^+/NADH system. NAD functions within the catabolic block, however, so that it is not seen in the diagram.

It is essential to emphasize the difference between such compounds as ATP, NAD, and NADP on the one hand and the intermediates of metabolic pathways on the other. The latter compounds are what we will call linear metabolites; that is, intermediates in linear sequences. (In this context, the citrate cycle is a linear sequence.) The compound is made in order to be converted into another compound, and carbon atoms actually flow down the pathway. The adenine and pyridine nucleotides, in contrast, are merely cyclically phosphorylated or oxidized, and the main molecule remains intact through thousands or millions of cycles.

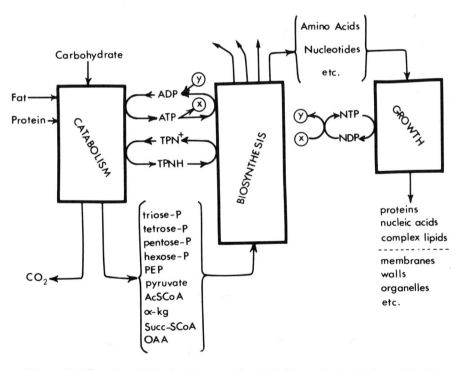

Figure 1. Functional block diagram of metabolism of a typical aerobic heterotrophic cell.

We will assume without arguing the point here that a system like the adenylate pool that picks up phosphoryl groups from about nine reactions and feeds phosphoryl groups— which in a metabolic context means energy—into several hundred reactions will not be able to function unless the ratio of ATP to ADP is held within very narrow limits. Even a relatively simple system such as a television set cannot function properly if the voltage changes very much. The interrelated sequences of reactions that collectively are life are far more complex than a television set. It is obvious that a small change in the ratio of ATP concentration to ADP concentration would be amplified enormously by the hundreds of reactions in which these compounds participate. Thus even in the absence of direct evidence, we should predict with a high level of confidence that the ATP/ADP ratio, or the mole fraction of ATP, would be maintained very nearly constant in a living cell.

Energy Charge

Individual enzymes probably respond primarily to concentration ratios. But because of the stoichiometric nature of the coupling functions of the adenine and pyridine nucleotides, a stoichiometric parameter seems more suitable in discussion of metabolism and regulation than does a ratio. In a two-component system such as $NADP^+/NADPH$ or $NAD^+/NADH$, the mole fraction itself is the appropriate parameter. Because the adenylate pool contains three components—ATP, ADP, and AMP—a simple mole fraction cannot adequately reflect the energy status of the pool. What is needed is the metabolically effective mole fraction of ATP. Adenylate kinase, which is found in all cells, catalyzes the reaction $2\ ADP \rightleftarrows AMP + ATP$. Thus two moles of ADP are equivalent, in terms of metabolic energy, to one of ATP. It then follows directly that the desired parameter expressing the energy status of the adenylate pool is the mole fraction of ATP plus half the mole fraction of ADP: $(ATP + \frac{1}{2}\ ADP)/(ATP + ADP + AMP)$. This parameter is the effective mole fraction of ATP; like any mole fraction, it may vary from 0 to 1. Because of its close correspondence to the charge of a storage battery, the effective ATP mole fraction has been termed the *energy charge* of the adenylate pool. As a consequence of the unique and ubiquitous role of the pool in metabolic energy transductions, the adenylate energy charge is a fundamental measure of the current energy status of a cell.

I commented earlier that the ratio of ATP concentration to ADP concentration must remain nearly constant if the adenylate pool is to be able effectively to fill its role as the ubiquitous energy transducing system. That is equivalent to saying that the value of the adenylate energy charge must be held within a narrow range in metabolizing cells. This has been found to be true. Energy charge values calculated from published analyses for ATP, ADP, and AMP in living cells of a wide variety of types were compiled in a 1971 paper (Chapman *et al.*, 1971). From these values and others published since, it seems clear that cells of most, and probably all, types maintain an energy charge of 0.85 to 0.95 when metabolizing and functioning normally. In earlier years there was a wide scatter, but as methods for quick sampling of tissues and cultures have improved, the results obtained have nearly all corresponded to energy charges very near 0.9.

In an electrochemical cell, as in many similar circumstances, the absolute concentrations of the reactants are relatively unimportant; the ratio of the reactant to product is the critical parameter. It seems intuitively obvious, in view of the nature of the coupling functions of the adenine and pyridine nucleotides, that in these cases also concentration ratios should be important. We might reasonably expect that organisms would have evolved mechanisms to maintain these ratios (or the adenylate energy charge) constant even when the absolute concentrations of the nucleotides vary. Several types of experiments have shown this expectation to be correct. Only one will be discussed here.

A chemostat, in which the volume of a bacterial culture is held constant by pumping in medium and removing culture at the same rate, provides great flexibility in the study of nutritionally limited growth. By appropriate manipulations of the pumping rate and of the composition of the medium added, the experimenter can obtain, in a time-independent state, any desired degree of growth limitation due to deprivation of any chosen essential nutrient. We have recently (Schwedes, Sedo, and Atkinson, *J. Biol. Chem.*, in press) grown an adenine-requiring strain of *Escherichia coli* in a chemostat at a wide range of growth rates with adenine as the limiting nutrient. A decrease in intracellular ATP level (and also in total adenine nucleotide pool) of 40% caused only a small decrease in growth rate, and the cells were able to grow slowly even after a 70% decrease in ATP concentration. At all growth rates, however, the energy charge was constant within the accuracy of the measurements (2 or 3%). This result clearly indicates that mechanisms have evolved for maintaining the energy charge within the narrow range that is compatible with growth even when the absolute concentrations of ATP, ADP, and AMP vary widely.

Some aspects of the molecular basis of this behavior have been observed earlier. In 1968 we reported (Atkinson, 1968) that several enzymes that participate in ATP-regenerating sequences responded to variation in energy charge as shown by curve R in Figure 2, whereas enzymes in sequences that utilize ATP responded as shown by curve U. Enough additional enzymes of both types have been studied to support a high degree of confidence that the pattern shown in the figure is a general one. Such responses clearly contribute to regulating the rates of metabolic sequences as appropriate—when energy is in good supply, as indicated by a high value of energy charge, biosyntheses and other ATP-requiring processes are favored, whereas a fall in the energy charge causes a sharp decrease in the rates of such sequences. Pathways in which ATP is regenerated also show the appropriate response—when ATP is needed, as indicated by a low value of energy charge, fluxes through these pathways are high. With increasing energy charge, the fluxes decrease. Interaction of responses of these two types must obviously contribute to stabilization of the energy charge, since a tendency for this value to drop will be opposed both by an increase in the rate of ATP production and by a decrease in the rate at which it is used. A tendency for the energy charge to rise would be opposed by oppositely-directed responses.

Figure 2 shows reaction velocites as a function of energy charge, and in fact many of our experiments have been done in this way; that is, we have measured the velocity of

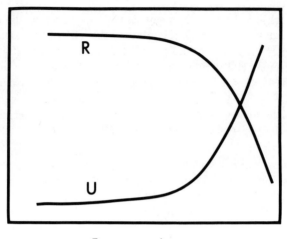

Energy charge

Figure 2. Generalized response to the energy charge of enzymes involved in regulation of ATP-regenerating (R) and ATP-utilizing (U) sequences. From Atkinson (1968).

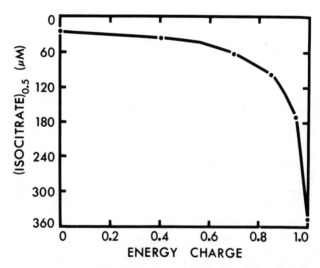

Figure 3. Concentration of isocitrate required for half-maximal velocity of the reaction catalyzed by yeast isocitrate dehydrogenase as a function of the energy charge. Reaction mixtures contained 100 mM Hepes-KOH (pH 7.6), 5 mM $MgSO_4$, 3 mM dithiothreitol, 0.3 mM isocitrate, and enzyme. The adenine nucleotide pool (ATP + ADP + AMP) was constant at 3 mM.

a reaction at fixed substrate concentration as a function of energy charge. As I discussed earlier, however, the primary effect of modifiers on regulatory enzymes is a change in the affinity of the catalytic site for the substrate. This is also true of responses to energy charge. Thus determination of such affinity as a function of energy charge gives a more operationally significant picture of an enzyme's regulatory properties than is given by measurement of velocity at a single substrate concentration. The most useful simple indication of affinity is the concentration of substrate required for half-maximal velocity. This parameter is sometimes termed the apparent Michaelis constant, but that name is not appropriate because the number is not at all constant; its variation in response to changes in concentrations of appropriate metabolites and in the value of the adenylate energy charge is the basis of metabolic regulation. Hence the symbol $S_{0.5}$ will be used here. The variation of $S_{0.5}$ for isocitrate of yeast isocitrate dehydrogenase is plotted as a function of energy charge in Figure 3. In such experiments, ATP, ADP, and AMP are added to a series of cuvettes at concentrations calculated to produce the desired values of energy charge. The components of the enzyme assay are then introduced and the initial velocity of the reaction is determined. When $S_{0.5}$ values are to be determined it is of course necessary to use a range of substrate concentrations at each value of energy charge. In Figure 3 the $S_{0.5}$ scale is inverted numerically so that the top will correspond to high affinity and hence to greater competitive effectiveness or to higher rate at a fixed concentration of substrate. The sharp increase in affinity for substrate as the energy charge drops from near 1.0 to 0.8 shows that the ability of the enzyme to obtain isocitrate is a sensitive function of energy charge in the physiological range. It should be remembered that the adenine nucleotides are not participants in the reaction catalyzed by isocitrate dehydrogenase; thus there is no basis in the chemistry of the reaction for any response whatever to changes in their concentrations or in the energy charge. The enzyme responds to adenylates because such a response is useful to the cell in which the enzyme occurs. Like other regulatory responses, this one could not in any way be predicted from knowledge of the reaction catalyzed. No clues to the understanding of such responses are to be found in the chemistry or physics of the reactions; the responses can be understood only in terms of evolutionary design.

Substrates and Products

It might be easy to assume that the supply-and-demand curves of Figure 2 satisfactorily explain metabolic regulation at a fundamental level. But brief consideration of the aspects of metabolism emphasized by the block diagram of Figure 1 shows that this assumption could not be valid. Most of the metabolites needed as starting materials for biosynthesis are produced by glycolysis and the citrate cycle. Thus if glycolysis were regulated only by energy charge in the manner indicated by curve R of Figure 2, the production of these starting materials would be depressed when the energy charge is high. The consequence would be that biosynthetic starting materials would be in poor supply when the energy charge was high. Such a response would evidently be undesirable; when the energy charge is high the cell should be capable of rapid biosynthesis.

An engineer faced with the need to design controls for a system that serves two functions would of course provide for the system to respond to two inputs, one to indicate how well each need is satisfied. Interacting controls of this type were developed by organisms several billion years before the first engineer was evolved. The type of two-response pattern to be expected for a regulatory enzyme in glycolysis is shown in Figure 4. When the concentrations of biosynthetic intermediates are low (upper curve), glycolysis should proceed even if the energy charge is high. If the supply of these intermediates is adequate, however, (bottom curve) glycolysis should respond primarily to the need for ATP, as indicated by the value of the energy charge, and its rate should be sharply depressed when the energy charge is high. Such a response insures that glycolysis will proceed at a relatively rapid rate (if substrate is available) whenever the cell needs either ATP or biosynthetic starting materials, and will be inhibited severely only when both are in good supply. A pattern of this type has been observed, for example, with rabbit muscle phosphofructokinase (Shen *et al.,* 1968). Although not itself a biosynthetic starting material, citrate serves in this case as the indicator of availability of these materials. This specific feature of the pattern could not have been predicted, but it is easily understood in terms of function. When the concentration of citrate is within the normal range, so will be those of α-ketoglutarate, succinyl coenzyme A, and oxalacetate, which are important starting materials for a variety of syntheses. Citrate is also the precursor of cytoplasmic acetyl coenzyme A, another important starting material. Perhaps this strategic position of citrate in metabolism, together with its low reactivity relative to the intermediates, accounts for its regulatory role.

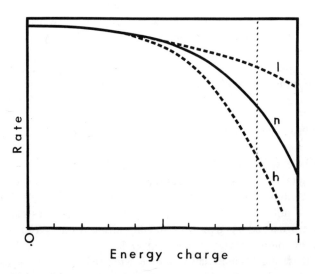

Figure 4. Generalized interaction between energy charge and the concentration of a metabolite modifier in the control of a regulatory enzyme in an amphibolic sequence. The curves correspond to low (1), normal (n), and high (h) concentrations of the metabolite modifier. From Atkinson (1968).

It is equally obvious that biosynthetic sequences would be unnecessarily wasteful if their rates were regulated only as shown by curve U of Figure 2, and all syntheses proceeded rapidly when the energy charge was high, whether the products were needed or not. Much more important than the resulting waste of ATP and starting materials, there would be no regulation of the concentrations of products. No cell could be expected to live in such a chaotic situation. Clearly the rate of each synthetic sequence must be independently regulated by the need for that synthesis, as indicated by the concentration of the end product. The expected pattern is seen in Figure 5. When the concentration of the end product is high (bottom curve), the rate of its synthesis should be low even if the energy charge is high. When the concentration of the product is low (upper curve), the cell should make more of it if its energy situation, as indicated by the value of the adenylate energy charge, is such that it can afford to do so. The effect of end product at an energy charge value of 1.0, seen along the right-hand vertical coordinate, is of course the negative feedback control of biosynthetic sequences that has been recognized since 1956. Interaction of this control with response to energy charge insures that a cell will make a specific metabolite only when that compound is needed and when the cell can afford the expenditure of energy that would be required for its synthesis.

Most regulatory enzymes probably respond to more than two inputs. Because the operational characteristic of the enzyme that is regulated is usually affinity for substrate, the concentration of substrate may also be considered a regulatory input, and of course it will at times become the primary limiting factor. Some enzymes are regulated

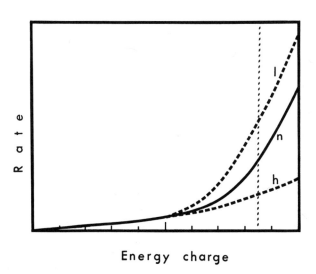

Energy charge

Figure 5. Generalized interaction between energy charge and end product concentration in the control of a regulatory enzyme in a biosynthetic sequence. The curves correspond to low (1), normal (n), and high (h) concentration of the end product modifier. From Atkinson (1968).

concurrently by the concentrations of many metabolites to whose syntheses they contribute. The biological advantages of other enzyme-modifier interactions may be less obvious but no less important. Phosphofructokinase, for example, is strongly stimulated by ammonium ion. This response may play an important part in carbon-nitrogen metabolism. When ammonium is available, glycolysis is favored, with consequent production of the starting materials for amino acid synthesis and hence for the production of protein. Conversely, when the supply of ammonia is limited, the rate of glycolysis will be depressed and gluconeogenesis, with production of storage carbohydrate, will be favored. Production of energy storage compounds, such as starch, when the nitrogen supply is limited has been recognized for many years.

Branchpoints

The most common type of regulatory interaction seems to be competition between two enzymes for a common substrate (a branchpoint metabolite), with the affinity for substrate of each of the competing enzymes being regulated by the adenylate energy charge and by one or more modifiers. Since the reactions of glycolysis, the pentose phosphate pathway, and the citrate cycle lead to the regeneration of ATP and NADPH but also provide the starting materials needed for biosynthesis, as illustrated in Figure 1, the question naturally arises whether there is a central branchpoint between degradative metabolism and biosynthesis. The answer is that interactions of metabolic pathways are too complex for a perfectly clean branchpoint between catabolic and anabolic pathways to be possible. During metabolism of carbohydrate by eukaryotic cells, however, pyruvate comes very close to being such a clean branchpoint.

There are two routes of entry into the citrate cycle from pyruvate: through acetyl coenzyme A and through oxalacetate. Entry of acetyl coenzyme A leads almost completely to oxidation, with regeneration of ATP. In one turn of the cycle, two molecules of CO_2 are produced, corresponding to the two atoms of carbon introduced as acetyl coenzyme A, and oxalacetate is regenerated. No intermediates are produced for use in biosynthesis. Entry of oxalacetate, on the other hand, can lead only to production of biosynthetic intermediates. Oxalacetate cannot be oxidized by the enzymes of the cycle except indirectly, by decarboxylation to three-carbon compounds, oxidation to acetyl coenzyme A, and re-entry in that form. (In speaking of the metabolic functions of acetyl coenzyme A and oxalacetate, we deal with metabolic *stoichiometry,* not with atom tracing. The entry of a molecule of acetyl coenzyme A into the citrate cycle is balanced by the production of two molecules of CO_2; it is of no significance in this connection that the actual carbon atoms lost are not, as shown by isotopic labelling, the same ones that were introduced at the beginning of that cycle.) When carbohydrates are being metabolized, the glycolytic pathway to pyruvate is both catabolic and anabolic. At pyruvate, as discussed above, a primary partition between catabolic and anabolic pathways occurs. This is prevented from being a clean separation only by the fact that one carbon atom of α-ketoglutarate, and thus of compounds derived from it, such as glutamate, glutamine, proline, hydroxyproline, and arginine, contain one carbon atom derived, stoichiometrically, from acetyl coenzyme A.

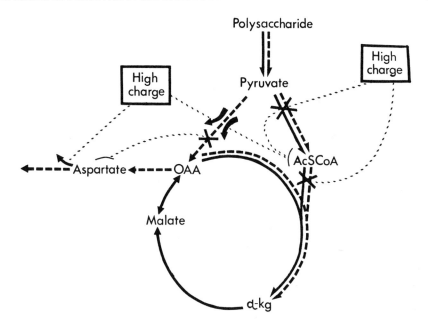

Figure 6. Some regulatory interactions at the pyruvate branchpoint in yeast grown on glucose. The broken arrows indicate biosynthetic sequences and the solid arrows indicate catabolic or energy-yielding pathways. Light broken lines connect modifiers to the reactions that they affect; a curved arrow indicates a positive effect and a cross a negative effect. "High charge" refers to a high value of the adenylate energy charge. From Miller and Atkinson (1972).

Control relationships at the pyruvate branchpoint in yeast growing on glucose are summarized in Figure 6. Solid arrows indicate catabolism leading to oxidation of product and regeneration of ATP, and broken arrows indicate pathways leading toward biosynthesis. Of the six regulatory interactions shown in this figure, four have been established in experiments using enzymes from yeast; the other two (the effects of high energy charge on pyruvate dehydrogenase and on aspartate kinase) were demonstrated with enzymes from *Escherichia coli* and are assumed to be valid also in yeast. At the branchpoint itself, where pyruvate carboxylase and the pyruvate dehydrogenase-decarboxylase complex compete for pyruvate, a high value of energy charge facilitates entry into synthetic metabolism by way of oxalacetate and decreases the tendency of carbon to be committed to degradation by conversion to acetyl coenzyme A. Negative feedback control of pyruvate carboxylase by aspartate should serve to adjust the rate of production of oxalacetate to metabolic needs for replenishment of amino acid pools. The direct effects of energy charge on the competing branchpoint enzymes will be supplemented by indirect effects. Thus an increase in energy charge causes a large increase in the $S_{0.5}$ value for acetyl coenzyme A of citrate synthase; the resulting increase in the concentration of acetyl coenzyme A will favor carboxylation of pyruvate. Similarly, if the response of yeast aspartate kinases to changes in energy charge value resembles that of the lysine-sensitive aspartate kinase of *E. coli*, an increase in charge will decrease the $S_{0.5}$ value of these enzymes for aspartate thus decreasing the concentration of aspartate and further favoring carboxylation of pyruvate.

The interactions shown in Figure 6 appear to provide for an appropriately-directed response to any conceivable fluctuation in concentrations or in metabolic needs affecting these reactions. For example, there are probably two major causes for a tendency for the concentration of acetyl coenzyme A to rise: a high value of energy charge, as discussed above, and a deficiency of oxalacetate. Both would decrease the rate at which acetyl coenzyme A is consumed in the citrate synthase reaction; high energy charge by decreasing the affinity of the enzyme for acetyl coenzyme A, and a low concentration of oxalacetate by substrate limitation. In both cases an increase in the rate of carboxylation of pyruvate is appropriate.

In many prokaryotes, the production of oxalacetate for entry into the citrate cycle occurs by carboxylation of phosphoenol pyruvate (PEP) rather than of pyruvate. In that case, PEP becomes the branchpoint metabolite, and the competing enzymes are PEP carboxylase and pyruvate kinase. We have similarly analyzed this branchpoint in *Azotobacter vinelandii* growing on sucrose (Liao and Atkinson, 1971). The situation is functionally nearly identical to that of the pyruvate branchpoint in eukaryotes.

When an organism uses a three-carbon compound as the sole source of carbon and energy, the catabolism-anabolism branchpoint again occurs at pyruvate, but because carbohydrates must now be made from three-carbon compounds rather than the reverse, the pattern is somewhat different. We analyzed interactions at this branchpoint in *E. coli* growing on lactate (Chulavatnatol and Atkinson, 1973a, 1973b). A key enzyme in this case is PEP synthase, which catalyzes the production of PEP, AMP, and orthophosphate from pyruvate and ATP. Because two phosphorylations are required to regenerate ATP from AMP, this reaction in effect uses two molecules of ATP. Thus production of PEP in this way is thermodynamically much more favorable than is production by reversal at the pyruvate kinase reaction. We found PEP synthase to respond positively to a high value of energy charge, as is appropriate for an enzyme catalyzing a reaction that commits the product to biosynthesis or gluconeogenesis. Pyruvate kinase responds oppositely; that is, it is an R-type enzyme in terms of the patterns shown in Figure 2. Of a wide variety of metabolites tested, PEP synthase was inhibited to an apparently significant degree by physiological concentrations of oxalacetate, PEP, α-ketoglutarate, L-malate, ADP-glucose, and 3-P-glyceraldehyde. These inhibitions are specific; no effect was observed with acetyl coenzyme A, NADPH, dihydroxyacetone phosphate, glucose-6-phosphate, fructose-6-phosphate, citrate, isocitrate, succinate, fumarate, 2-P-glycerate, L-lactate, L-aspartate, L-alanine, or L-phenylalanine. The functional reason for this pattern of inhibition is clear when a diagram of metabolic pathways in a lactate-grown cell is constructed (Figure 7). All of the compounds found to inhibit PEP synthase are produced from PEP, so their participation in negative feedback control of the rate of PEP

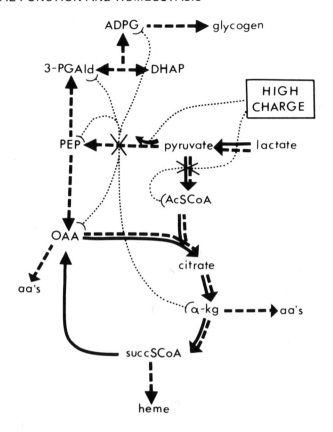

Figure 7. Some regulatory interactions at the pyruvate branchpoint in *Escherichia coli* grown on lactate. Significance of arrows and other symbols is as in Figure 6. From Chulavatnatol and Atkinson, 1973a.

production is clearly appropriate. Most of the inhibitory compounds occupy rather strategic positions in metabolism; for example, ADP-glucose is in *E. coli* the precursor of storage carbohydrate (Preiss, 1969). The inhibitory effect of malate is not shown in the figure for reasons of simplicity. Because of the unfavorable equilibrium of the malate dehydrogenase reaction, changes in the concentration of malate probably parallel those in oxalacetate concentration, and this response is probably a reinforcement of the inhibition by oxalacetate.

Enzyme-Substrate Affinities

The curves of Figure 2 suggest how the value of the energy charge can be stabilized if total concentrations remain constant, but in themselves they do not explain how the charge can be regulated in the face of large variations in the total concentration of adenine nucleotides. Again, the results of experiments on regulatory enzymes *in vitro* supply at least a partial answer. The different metabolic functions of the intermediates and products of linear metabolic sequences on the one hand and of metabolic coupling agents such as the adenine and pyridine nucleotides on the other is reflected in differences in the evolved behavior of enzymes. On the basis of current evidence, there appear to be two fundamental differences.

First, $S_{0.5}$ values for substrates that are intermediates in linear sequences are generally about equal to or somewhat larger than the physiological concentrations of the substrates. Therefore the catalytic site is only partially saturated with the substrate, and the rate of the reaction is sensitive to changes either in the concentration of substrate or in the $S_{0.5}$ value. In contrast, $S_{0.5}$ values of many kinases and dehydrogenases for ATP and ADP, for $NADP^+$ and NADPH, or for NAD^+ and NADH are much smaller than the physiological concentrations. In consequence, the sites at which these coupling agents bind are essentially saturated at all times, and the only change available is a shift in the ratio of the number of enzyme molecules binding ATP to the number binding ADP. This will of course depend on the concentration ratio of unbound ATP to unbound ADP and on the $S_{0.5}$ values for the two nucleotides. A similar pattern is found at the catalytic sites of many dehydrogenases. In such cases changes in total concentration of the coupling agents, within rather wide limits, will not affect the rate of the reaction. Because the catalytic site remains saturated, the enzyme cannot sense changes in total concentrations. The rate of reaction will, however, be sensitive to changes in the ratio of ATP to ADP (hence to changes in the value of energy charge) or of $NADP^+$ to NADPH or NAD^+ to NADH. This type of response has been observed *in vitro* with several kinases and dehydrogenases. Kinases giving energy charge response curves like curve U in Figure 2 show the same response when the total adenylate concentration is changed several-fold. Similar patterns have been observed for dehydrogenases, where the curve of reaction rate as a function, for example, of NAD^+ mole fraction is not changed by a several-fold increase or decrease in the total NAD^+-NADH pool. It should be noted that a single catalytic site binds both an intermediate in a metabolic sequence and the nucleotide coupling agent. Thus the distinction with regard to relative affinities occurs within a single catalytic site—the $S_{0.5}$ value for the substrate to be phosphorylated or oxidized will be larger than the physiological concentration of that substrate, while at the same catalytic site the $S_{0.5}$ values for the adenine or pyridine nucleotide coupling agents will be much smaller than the physiological concentrations of these compounds.

The relationships just described explain how response curves like those of Figure 2 can be independent, within limits, of changes in concentration of the coupling agents, but they do not explain how the shapes of the curves are determined. In the simplest cases, these depend on the relative affinities for the two alternate forms of the coupling agent—ATP and ADP in the case of kinases. Kinases will respond to variation in energy charge value as shown by curve U of Figure 2 if the affinity for the product nucleotide, ADP, is around 4 to 10 times greater than the affinity for the reactant, ATP. A similar pattern of response to the NAD^+ mole fraction is shown by many dehydrogenases that catalyze primary oxidations of substrates in degradative sequences. This pattern, too, results from greater affinity for the product nucleotide (in this case, NADH) than for the reactant (NAD^+). From the standpoint of catalytic efficiency, it would make no sense for an enzyme to bind a product more firmly than the corresponding reactant. The inevitable consequence must be a marked degree of product inhibition. But this product inhibition is turned to biological advantage in the adjustment of response curves for optimal metabolic benefit. This unexpected pattern of affinities is the second difference in the behavior of enzymes toward coupling agents as compared to their handling of regular metabolites in linear sequences. Together the two account for a highly advantageous phenotypic consequence—response curves like curve U of Figure 2 that are relatively independent of fluctuations in the pool of a coupling agent.

Interactions between substrate binding affinities are discussed in more detail elsewhere (Atkinson *et al.*, 1975). They are mentioned here to illustrate the point that many phenotypic characteristics seem, when considered alone, to make no sense at all. This is true of the properties of enzymes as of any other feature. When seen in context, however, very nearly all characteristics of an organism must have survival value. If the functional characteristics illustrated in Figure 1 were not taken into account, the strange properties of kinases and dehydrogenases with regard to the binding of adenine and pyridine nucleotides would seem strange and even counterproductive (since an enzyme that binds the product more firmly than a reactant is necessarily a less efficient catalyst than if the relative strengths of binding were reversed). But when the special metabolic roles of the coupling agents are considered, the observed binding patterns are seen to be essential to proper metabolic control and thus to be a necessary condition for life.

Enzymes and Evolution

This example, from the field of my own research interests, illustrates the very general concept that many properties of enzymes, other components of organisms, or of the intact organisms themselves which are meaningless or even apparently harmful when considered individually are seen to be functionally advantageous when their operational interactions with other enzymes, other components, or other properties are considered. The principles of design are similar whether the designed object be helicopter, yeast cell, machine tool, or man. Only in the last few years, however, have we been in a position to begin asking questions concerning functional design at the molecular level. We can, I think, begin to supply some generalizations that will be of broad applicability in biology. It is very important, however, for those of us who are biochemists or molecular biologists to remember that biology does not begin and end at the molecular level. It is a truism, but one sometimes overlooked, that mutation is at the molecular level but selection is at the level of the entire organism. It seems to me that a distinction that one very frequently finds made in general articles about biology and especially about evolutionary biology— the distinction between functional and molecular biology on the one hand and evolutionary biology on the other—is not only a distinction without a difference but is highly undesirable and divisive. We could not have evolution without function; function is what is selected; but I hope that it is equally clear that we could not have function without evolution. Evolution is the design of function. It is evident, I think, that biology is just now entering its most exciting period with regard to basic generalizations. We know the genetic alphabet. We know some generalizations regarding metabolic control. In nearly every area of biology, from molecular genetics to animal behavior, we now have an inkling of some underlying generalizations, but in no field have we more than a primitive beginning of real understanding of fundamental relationships. In spite of that I think that it is not unreasonable, although perhaps a bit parochial, to suggest that advances in biology in the last fifty years constitute one of the great intellectual achievements of mankind, perhaps second only to development of language, which doubtless took considerably longer. Advances during the next fifty years will be much more extensive, at least if society sees fit to continue support of fundamental biological research. The study of biology is now and will continue to be one of the most important and exciting adventures in human history. We can confidently hope and expect that the new School of Life Sciences whose birth we are commemorating in this symposium will participate in the excitement and contribute to the advances in biological research for many years to come.

Acknowledgment

I wish to take the opportunity, at this University of Nebraska symposium, to acknowledge my indebtedness to Dr. L.C. Newell of the University of Nebraska and the U.S. Department of Agriculture, with whom I was privileged to work as an undergraduate. My nearly daily contacts with Dr. Newell provided my first interaction with a working scientist, and his simultaneous interest in broad general questions and in practical agronomic problems strongly influenced my scientific development.

REFERENCES

Atkinson, D.E. (1968). *Biochemistry 7,* 4030.

Atkinson, D.E., Roach, P.J. and Schwedes, J.S. (1975). *Adv. Enzyme Regulation 13.* (In press.)

Barnes, L.D., McGuire, J.J. and Atkinson, D.E. (1972). *Biochemistry 11,* 4322.

Chapman, A.G., Fall, L. and Atkinson, D.E. (1971). *J. Bacteriol. 108,* 1072.

Chulavatnatol, M. and Atkinson, D.E. (1973a). *J. Biol. Chem. 248,* 2712.

Chulavatnatol, M. and Atkinson, D.E. (1973b). *J. Biol. Chem. 248,* 2716.

Liao, C.L. and Atkinson, D.E. (1971). *J. Bacteriol. 106,* 37.

Miller, A.L. and Atkinson, D.E. (1972). *Arch. Biochem. Biophys. 152,* 531.

Preiss, J. (1969). *Current Topics in Cellular Regulation,* eds. Horecker, B.L. and Stadtman, E.R. (New York and London: Academic Press), Volume 1, 125.

Shen, L.C., Fall, L., Walton, G.M. and Atkinson, D.E. (1968). *Biochemistry 7,* 4041.

SOMATIC CELL GENETICS AND ITS HUMAN APPLICATIONS

Theodore T. Puck

Department of Biophysics and Genetics
*University of Colorado Medical Center**
Denver, Colorado

Introduction

Until recently, biology was a science which dealt mainly with observation and description of living forms. Less than twenty years ago a profound revolution occurred which transformed it into a powerful theoretical science as well. This new theoretical base has enormously extended the scope of biology's operations and its ability to understand, predict and control phenomena at many different levels. The new understanding of the nature of the living process came about from the contributions from three different kinds of fields, genetics, biochemistry and microbiology.

These studies uncovered the chemical structure of the gene, the ultimate unit of heredity by which parents transmit the potentialities for specific characteristics to their offspring. Every living cell contains a set of genes which are the ultimate units of information by which cell structures are built and their functions regulated. The genes act in two ways. They duplicate themselves, a function which is the true basis of biological reproduction. In addition, they form the blueprints which specify the synthesis of the proteins which are the molecules that constitute the basic machinery of the cell. Some genes are regulatory, *i.e.*, they produce substances that turn off or turn on other gene sets, so that each cell makes only those proteins which it needs at any particular time. The proteins that are present in the cell have their activities regulated by a complex set of feedback operations so that at each instant in time the cell's chemical activities are exquisitely adjusted for maximum effectiveness in the particular physical and molecular environment surrounding the cell.

*This investigation is a contribution from the Eleanor Roosevelt Institute for Cancer Research and the Department of Biophysics and Genetics (Number 494), University of Colorado Medical Center, Denver, Colorado.

Not only did the new molecular biology furnish a model of how a cell actually works but also provided a picture of the molecular nature of evolutionary processes. When the first self-duplicating macromolecule appeared on the face of the earth its reproduction was accomplished by incorporating into a structure identical to itself the necessary small molecules which happened to become available in the surrounding environment. Because of the intrinsic uncertainties in molecular behavior, sooner or later a changed form of the self-replicating molecule had to arise. At this point a competition would inevitably be set up between this form and the parental form. Each would compete in the capture of the small molecules needed for the duplicating processes. In such a system one of the competing forms would have some advantage, however slight, over the other. Therefore, it would replicate itself at the expense of the other and its numbers would grow exponentially. The nature of exponential growth is such that very soon the less competent form would be swamped out and would vanish from the scene. But then the process inevitably would repeat itself. Sooner or later altered forms of the new reproducing form would appear and again a competition would result. During the course of many, many millions of years, this competition would result in living forms more and more successful in their ability to capture the small molecules needed for their own duplication. As a result of this never-ending competition, the surviving forms became ever more efficient in their ability to manipulate molecules found in their environment. Nowhere in the universe except in living systems does there exist the ability to order molecular behavior so effectively. The later introduction of sex into biological reproduction caused a great increase in the rate of evolution because it permitted many new combinations of self-replicating or genetic structures to be tested in this continuing game of self-selection. Thus given the process of self-duplication of giant molecules from smaller molecules available in limited amounts in the environment, the vast array of living behavior as we know it could have arisen, introducing an astonishing degree of order in an otherwise disordered and random universe.

New Approaches to Human Genetics

Molecular biology arose through the fusion of biochemical and genetic studies in microorganisms. It was natural to expect that these fundamental advances in understanding would also produce new information about human biology. However, a large stumbling block lay in the path of such application due to the difficulty in acquiring the needed human genetic information. Microorganisms can reproduce in approximately 25 minutes whereas the generation time of man is about 25 years. Moreover, it is not possible in the case of man, to carry out those matings which would be most illuminating from the genetic point of view. The classical methods for studying human genetics which involved approaches like pedigree analysis, appeared to be insufficiently powerful for application of the new molecular biology.

A simple solution to the problem occurred as a result of the following considerations. Man and other mammals are after all made up of individual cells each of which resembles, at least in a formal sense, the cells of microorganisms. Why not then, take samples of these cells from the skin or any other tissue of a mammal, grow these outside the body by methods of tissue culture until large populations are obtained, and then study the genetics of these cells by the very methods which had previously been developed for study of microorganisms.   The principal operation that was required was to be able to grow large numbers of isolated single cells into individual colonies. This permits mutants to be recognized, and to be picked for subsequent genetic analysis. If one could grow single mammalian cells in this way, one could study genetics and genetic biochemistry at the cellular level, and hence bypass the need for reproduction of the whole mammalian organism.   The mechanical arrangements necessary to produce the desired objective proved astonishingly simple and, in a short time, it proved possible to place single cells from almost any mammal into glass dishes, allow these to grow into large, readily visible colonies of cells, and to carry out the kinds of operations with such colonies that have given birth to the science of microbial genetics and molecular biology. The results of these operations have had tremendous impact in enlarging our understanding of human genetics and genetic biochemistry. The use of cell growth outside the body permitted two basic kinds of operation in the study of mammalian cell behavior. First, one could study mutations which act as markers to permit tracing of hereditary processes and definition of the genetic structures. Second, it became possible to carry out precise measurements of mammalian cell reproduction. Thus after depositing a number of single cells in a dish one could count the number of colonies developing from them. The "plating efficiency," which is the ratio of the number of colonies produced to the number of single cells deposited in a standard dish has come to be a precise method for characterizing physical, chemical and biological agents which are capable of modifying the growth of mammalian cells. Such measurements of cell growth are important in the study of cancer, the nature of drug action, and the effect of physical agents like x-rays and ultraviolet light on mammalian cells.   In this way it is possible to substitute precise numerical values for subjective impressions in determining how different agents effect mammalian cell reproduction.

The Human Chromosomes and Human Chromosomal Diseases

The genetics of an animal like man is far more complex than that of the simplest microorganism like bacteria.   Each human cell contains almost a thousand times more gene material than does a typical bacterium. Genes are contained on a linear chain, called a chromosome. A bacterium like *E. coli* has only one chromosome, but mammals have about 20 to 30 pairs of chromosomes. There is a special pair of sex chromosomes which determines whether the individual shall be male or female. All of the other chromosomes are called autosomes. Thus, the genetics of man must be studied at two different levels, that of the chromosome as well as of that of the individual genes.

For over thirty years the human chromosomes had constituted a dark continent of science. Not even their correct number was known. With the advent of the new techniques for studying normal mammalian cells outside of the body, the situation became transformed. Hand-in-hand with the new scientific understanding there has emerged an astonishing variety of application to problems of human health and disease, and far-reaching implications for man's future.

Within a remarkably short time, techniques were developed for making the human chromosomes visible. In 1959 an international study group developed the classification and numbering system which is used today. While everyone expected that there would eventually be important medical applications, no one was prepared for the large number of these and their deep implications for the biology of man. Routine methods for examination of the chromosomes of any person were developed and the chromosomes of large numbers of healthy and diseased persons were studied. As a result, some astonishing facts emerged.

The normal human chromosome number has been established as 46 there being 22 pairs of autosomes and one pair of sex chromosomes. In the female both of these are the same and are called X chromosomes. The male has one X and one much smaller chromosome called the Y. All the cells of the body except the reproductive cells have this same chromosomal constitution. Any deviation from the normal chromosome number is attended by the most deep seated disease, although in some cases the disease situation may not manifest itself until a future generation. Many of the diseases which have constituted classical enigmas of medicine have now been identified as manifestations of chromosomal aberration. These include Down's Syndrome or Mongolism, Turner's Syndrome, a condition in which the newborn is a female but never develops sexual maturity, and a host of other human defects. The amazing fact has emerged that approximately one half to 1% of all human live births produce children with chromosomal aberrations who are destined to develop severe abnormalities. The majority of the diseases so produced involve mental deficiency as one of the accompanying features. These studies brought to light that chromosomal defects constitute a major public health problem which appears to affect human populations at a wide variety of different economic levels.

These diseases can be produced by errors in structure as well as in the number of chromosomes. Thus Mongolism which normally occurs as a result of the presence of an extra chromosome number 21, can also be produced if only an extra piece of this chromosome is present, attached to another chromosome. Similarly, deletion of a small piece of chromosome number 22 in cells of the bone marrow can produce one form of leukemia.

As so often happens in science, an understanding of the dynamics of these problem situations soon opened up possibilities for their control. Ideally, one would like to prevent production of these chromosomal aberrations or find means to bypass their deleterious effects. However, until these powers are achieved, another avenue appears feasible. If one could diagnose the presence of these diseases *in utero,* in time to terminate the pregnancy safely, one could hope to eliminate births of seriously handicapped children with chromosomal aberrations. This last possibility has been the easiest one to achieve and at the present time can be routinely applied as a monitoring system to scan human births. The scientific and technological means are now available to prevent the birth of children with any one of approximately 40 different chromosomal diseases. It is necessary simply to take a small sample of the fluid surrounding the fetus during the 12th to the 16th week of pregnancy. This fluid contains fetal cells. These cells can be deposited in glass dishes in a nutrient medium which permits them to reproduce and form colonies. The chromosomal constitution of these cells can then be examined and the presence of chromosomal disease can be diagnosed with a high degree of certainty.

These procedures make it possible to prevent birth of children with a large variety of terrible diseases. Families need no longer fear such tragedies. Moreover, by use of this technique for monitoring the foetus *in utero,* families with a high risk for production of children with genetic defects of this kind can almost always now be assured of a baby free from these particular diseases. Pregnancy can be initiated and the fetal cells examined. Pregnancies which involve chromosomal defects are terminated, and the parents can initiate a new pregnancy. Only normal embryos are permitted to proceed to maturation. When one considers the terrible psychological effects on the families of children born with devastating and incurable defects, and the cost to society of maintaining such patients, the full implications of these new developments can be appreciated. By far the greater part of the cost of human chromosomal disease is not to be measured only in dollars and cents. However, even this most crude and incomplete measurement of this burden is illuminating. It has been estimated that the United States spends 1.8 billion dollars a year to care for its population of patients affected with Mongolism. This represents only one of the 40 odd chromosomal diseases which are now preventable.

These new developments require changes in our moral and legal attitudes. It would seem to be immoral to permit the preventable birth of a severely defective human child, particularly when the scientific basis exists for permitting the family involved to give birth to a normal child instead. These new powers make it necessary to change some of our traditional attitudes. It seems reasonable now to propose that each child born is entitled to the best genetic constitution that science and technology can provide. Each child born would also appear entitled to receive a developmental history which would offer the greatest probability for fulfillment of the potential of his genetic constitution. While the fulfillment of genetic potentialities always involve strong modulation by social values, care must be exercised to permit the widest possible variety of normal human genetic differences. These new powers must not be used to coerce or force conformation to arbitrary standards. But parents who wish to use these new possibilities to escape the threat of grossly defective children must be enabled to do so.

Gene Mutations

Chromosomal aberrations are only one of the genetic defects to which man is susceptible. What about gene mutations which can also produce severe and pitiful defects in human beings? This region of research has also expanded enormously through the study of mammalian cells cultivated outside the body. It has become possible experimentally to produce mutations in such cells, to identify and isolate the mutant cells, and to study the biochemical defects associated with such mutations. Such studies have made it possible to screen the action of new drugs, food additives and environmental pollutants for their mutagenic action on mammalian cells. Presumably it will now be possible to protect human populations from a variety of genetic insults to which our industrial culture might otherwise subject them.

Just as in the case of chromosomal aberrations, many gene mutations can also be diagnosed *in utero.* It has been estimated that the number of severe, single gene diseases that could be so diagnosed is about equal to the number of chromosomal defects which can be similarly recognized. One example of a single-gene defect which can be recognized by this technique is the Lesch-Nyhan Syndrome. This condition is particularly interesting because it is the first of the cerebral palsy group of diseases to have its biochemistry identified. This condition is particularly heartbreaking because the afflicted children are driven to mutilate themselves in a most pitiful fashion.

Cell Hybridization and Gene Dominance

Studies of mammalian cell genetics outside of the body have been enormously aided by the phenomenon of cell hybridization. This is a procedure in which two cells coalesce to form a new composite cell containing the combined genetic structures of the two partners. It permits one to carry out many of the operations of genetic analysis for which sexual mating between entire organisms was heretofore needed. For example, one can easily determine whether a given form of a gene is dominant or recessive with respect to an alternate form. Two cells, each with different forms of a particular gene are hybridized. One then observes which characteristic develops in the colonies resulting from the multiplication of the composite cell. The determination of dominance is one of the classical operations which required mating between two complete organisms. The ability to carry out such a procedure rapidly and simply, using human cells in glass vessels instead of studying the results of human matings demonstrates the power of this new approach to human genetics.

Gene Mapping

These new methodologies have lent themselves to determination of the chromosomal location of the various human genes. Why should one want to map the human genes? First, because if we knew which genes lie on which chromosomes then we would know which biochemical activities are affected in patients with chromosomal aberrations. Such information is essential if we are ever to understand the biochemistry of the resulting disease and attempt to design approaches to cure or relieve the condition. Furthermore, if we had a simple biochemical test that would identify the presence of extra or missing chromosomes, the process of diagnosis of chromosomal disease might be greatly simplified. At present, the cost of a complete analysis of the chromosomes of a human patient is approximately $100, a large amount of money for most families.

There is still another important reason why we must map the human genes. The central problem of human biology in our time involves the need to understand how particular genes are turned on or turned off in particular cells at different times. This is the process which underlies the mechanism of differentiation. In mammals a single fertilized egg reproduces to form a variety of cells with different structures and biochemical activities. Thus the original single cell from which each human life begins, produces populations of daughter cells that become heart, liver, lung and skin, each containing cells with different structures and biochemical activities. As noted earlier, it has been demonstrated that the chromosomes of all the body cells are identical. Therefore, all cells contain all of the genes, but specific mechanisms must operate so that in each tissue only those genes are functional which are required for its specific actions. Thus genes for hemoglobin synthesis, for example, are active only in the blood-forming cells, but are dormant in the cells of the rest of the body.

The only good model we have for gene regulation in cells comes from studies in bacteria. In these microorganisms, it has been found that genes adjacent to each other form groups which are simultaneously turned on or turned off in accordance with the needs of the cell. Do similar kinds of regulatory gene groups exist in mammalian cells? To answer this question it is necessary to map the genes so as to find out whether genes which are turned on and off together do indeed lie next to each other on a chromosome. An understanding of these regulatory mechanisms should clarify many problems ranging from embryonic development to cancer.

Exploration of the Mechanism of Action of Drugs

These new techniques open up new and powerful approaches to understanding the mode of action of drugs and chemical agents on mammalian cells. The cellular action of only an extremely small fraction of the drugs listed in the Pharmacopoeia is understood. It is now possible to measure accurately the effect of any drug on the following processes: (a) initiation or suppression of cell multiplication, (b) alteration of the rate of cell multiplication, (c) change in the molecular requirements of the cell for multiplication, (d) pin-pointing the place in the cell life cycle where the drug operates, (e) testing which drugs antagonize or augment each other's action at the cellular level, and (f) effect of the drug on synthesis of particular molecules by the cell. These approaches may be expected not only to explain how currently known drugs operate, but to provide powerful screening techniques by which new drugs with specific actions can be discovered.

The Prospect for the Future

The molecular biology of the mammalian cell is offering many new pathways to attack problems of development and differentiation and is opening new approaches to understanding of cellular coordinating activities like those which go on through the mediation of the brain and the hormones. Until recently, the action of hormones at the cellular level was highly mysterious. Now several new leads have become available. It has been shown that hormones acting on specific mammalian cells can induce synthesis of new enzymes in the affected cells. Many hormones have been demonstrated to act via a common mediator molecule, cyclic adenosine monophosphate (cyclic AMP) which is formed inside the cell in response to addition of certain hormones. Cell mutants have been isolated which require specific hormones in order to grow, and others have been developed which will synthesize large amounts of specific hormones in tissue culture. It has recently been demonstrated that cyclic AMP, particularly when combined with certain hormones, can cause cells to lose characteristics which are associated with cancer. These various experimental approaches offer great intellectual adventure to the scientist and the possibility of acquiring enormous new powers in areas of great human problems.

Application of the new techniques of cellular and molecular biology to man offers a wide variety of new approaches to an understanding of development. The whole battery of biochemical and genetic tools here described are ready now to be applied to these problems and should furnish enormously powerful information. But other techniques also suggest themselves. For example, it is now possible to identify at birth babies with defects in specific chromosomes. Many of these children appear clinically normal at birth. Yet by the time adulthood has been reached, these individuals often show characteristic defects both in body structures and in behavioral patterns. Somewhere between birth and adulthood certain normal processes have failed to occur in this particular group of patients. By following these patients carefully throughout their development, there is hope of observing the point at which they first exhibit deviations from normal developmental processes. Studies like these, which are now in progress, may help to pinpoint reactions which constitute part of normal developmental processes but which are blocked in these chromosomally defective persons. This approach involves applying to man one of the classical kinds of biological analysis, namely using mutants to unravel steps in normal metabolism which are otherwise difficult to isolate.

Study of such defective developmental patterns in persons with defects in the sex chromosomes also promises to furnish illumination about what aspects of human sexual behavior are genetically determined and which are primarily cultural in their expression. The degree to which genetic and cultural elements contribute to the development of human behavioral patterns may constitute one of the most important scientific problems in modern human biology.

The new developments in the science of man offer hope of clarification of some of the classical problems concerning human behavioral needs and drive which have plagued poets, philosophers and religionists for thousands of years. Recent studies with a variety of animals have uncovered the existence of genetically determined drives which affect aggressiveness and hostility patterns. Similarly the formation of various kinds of social bonds between individuals in animal societies is strongly influenced by genetic patterns. One example is the "imprinting reaction," whereby ducklings can be induced to accept as their mother any figure which is presented to them in an appropriate manner within a specified time after hatching. Similarly, baby monkeys have been shown to require certain kinds of mothering in the earliest periods of infancy if they are to grow into normal adults.

It becomes of critical importance to understand the nature of the basic genetic-behavioral drives in man, the biochemical modes involved in their expression, and the way in which expression of these drives can be modified by social and cultural influences. The complexity of the modern world has made the traditional kinds of value systems which in simpler societies have served to provide satisfactory goals and regulators of human behavior no longer adequate. The birth of the new science of man offers hope of establishing bridges between the simpler kinds of genetic determinants and the more complex genetic behavioral patterns. Understanding of these dynamics should furnish man with new and powerful tools to build societies free of the devastation that humans have traditionally inflicted on one another and which can promote human fulfillment in an intensity and breadth undreamed of in previous periods of human history.

The newly-emergent science of man which appears on our horizon will provide extremely powerful new tools governing human behavior at many different levels. These, like any new powers, are in themselves neutral and can be used equally effectively for human benefit or degradation. If these powers are to be used to increase human fulfillment, there must be a concerted program undertaken to educate society into the meaning and possibilities of these new powers. While man is a mammal and shares many of the characteristics of other animals, he also possesses characteristics and potentialities which are unique and which require carefully controlled developmental history if these are to develop to their point of maximum fulfillment. The new powers would make it possible on the one hand either to enslave man more effectively than ever before or to create a golden age for all humanity. Never before was such a choice available. Achievement of these new possibilities for human fulfillment, which include a peaceful and uncontaminated world, represents the true challenge of our new age.

INDEX